Springer Geography

The Springer Geography series seeks to publish a broad portfolio of scientific books, aiming at researchers, students, and everyone interested in geographical research. The series includes peer-reviewed monographs, edited volumes, textbooks, and conference proceedings. It covers the major topics in geography and geographical sciences including, but not limited to; Economic Geography, Landscape and Urban Planning, Urban Geography, Physical Geography and Environmental Geography.

Springer Geography—now indexed in Scopus.

More information about this series at http://www.springer.com/series/10180

Courage Kamusoko

Remote Sensing Image Classification in R

Springer

Courage Kamusoko
Asia Air Survey Co., Ltd.
Kawasaki, Kanagawa, Japan

ISSN 2194-315X ISSN 2194-3168 (electronic)
Springer Geography
ISBN 978-981-13-8011-2 ISBN 978-981-13-8012-9 (eBook)
https://doi.org/10.1007/978-981-13-8012-9

This Springer imprint is published by the registered company Springer Nature Singapore Pte Ltd.
The registered company address is: 152 Beach Road, #21-01/04 Gateway East, Singapore 189721, Singapore

To my girls Ai and Reina, who provide unending inspiration!!

Preface

This workbook is an introduction to remote sensing image processing and classification in R. The decision to write this workbook is based on my experience during my graduate and post-graduate studies, university teaching, and work as a consultant in remote sensing and modeling land use/cover changes. The aim of the workbook is to assist people who cannot afford commercial software and those who want to implement machine learning classifiers for remote sensing image processing and classification in R. There are many excellent introductory and advanced books on R, and remote sensing. However, there are few books that guide students, researchers, university teachers or other remote sensing practitioners for practical implementation of remote sensing in R. Therefore, this workbook is designed to be concise and practical in nature not as a complete guide to image processing and classification in R. That is, it is more of desktop reference workbook, which introduces R so that one can immediately start using the software platform and R packages for image processing and classification. Although R has an initial step learning curve, it is worth investing in R because it is free and has many packages, which might not be available in commercial software.

I also want to make it clear that while R is a good software or platform for graphics, statistical analysis, and machine learning given the number of packages available, it might not be appropriate for the analysis of massive remotely-sensed data. This is because the data being analyzed in R is held in memory. However, there many innovations in big data analytics such as parallel computing technologies in R or even cloud computing. Alternatively, other programming or scripting languages such as Python, Perl, and Ruby can be used depending on the nature of the problem and data availability.

Who should use this workbook?

The workbook is for undergraduate and graduate students in remote sensing and geographic information science or other related disciplines. While I assume no prior knowledge of R, the basic understanding of remote sensing is required. The workbook is also aimed at university teaching staff, researchers, or anyone interested in remote sensing image processing and classification. In addition, consultants or other people who are familiar with remote sensing but have limited experience in R can use this book to quickly test machine learning classifiers on small data sets.

How is this workbook organized?

This workbook is organized into five chapters. Chapter 1 introduces remote sensing digital image processing in R, which is subdivided into six sections. Section 1.1 presents a brief background on remote sensing image processing and classification, while Sect. 1.2 provides a brief overview of R and RStudio. Section 1.3 describes the data and test site, while Sect. 1.4 provides a quick guide to R. Finally, Sects. 1.5 and 1.6 provide the summary and additional exercises.

Chapter 2 covers pre-processing. Section 2.1 provides a brief background on pre-processing. Next, Sects. 2.2 and 2.3 provide tutorial exercises 1 and 2, which focus on displaying Landsat 5 Thematic Mapper (TM) imagery, and radiometric correction and reprojection. Finally, Sects. 2.4 and 2.5 provide the summary and additional exercises.

Chapter 3 focuses on image transformation. Section 3.1 provides a brief background on image transformation. Next, Sects. 3.2 and 3.3 focus on computing vegetation and texture indices. The final Sects. 3.4 and 3.5 provide the summary and additional exercises, respectively.

Chapter 4 covers image classification. Sect. 4.1 provides an overview of remote sensing image classification. Next, Sects. 4.2 and 4.3

focus on single date and multidate image classification, respectively. Finally, Sects. 4.4 and 4.5 provide the summary and additional exercises, respectively.

Last but not least, Chap. 5 focuses on improving image classification. Section 5.1 presents an overview of improving image classification, while Sect. 5.2 provides a brief background on feature selection. Next Sect. 5.3 focuses on image classification using multiple predictor variables, while Sect. 5.4 focuses on image classification using multiple predictor variables and feature selection. Finally, Sects. 5.5 and 5.6 provide the summary and additional exercises. An attempt has been made to organize the workbook in a general sequence of topics. Therefore, I encourage you to read the workbook in sequence from Chap. 1.

Conventions used in this workbook

The R commands or scripts are written in Lucida Console font size 10 in italics, while the output from the R is written in Lucida Console font size 10. Note that long output from R code is omitted from the workbook to save space. In some cases, I use small font sizes in Lucida Console to show how the R output or results would appear. This is just for illustration purposes. Readers will of course see the whole R output when they execute the commands. The hash sign (#) at the start of a line of code indicates that it is a comment. Finally, all explanations are written in New Times Roman font size 12.

Data sets, R scripts, and online resources

Information about the data sets and some sample scripts used in this workbook are available. Furthermore, I provide additional online resources on R, R packages or remote sensing in the appendix.

Kawasaki, Japan Courage Kamusoko

Acknowledgements

First and foremost, I would like to express my sincere gratitude to Professor Yuji Murayama, who has been inviting me to deliver a "Special Lecture on Geoenvironmental Sciences", with a focus on "Practical Machine Learning for Geospatial Processing and Modeling" at the Graduate School of Life and Environmental Sciences, University of Tsukuba in Japan. This workbook would not have been possible without his support. Second, I would like to thank students from the Graduate School of Life and Environmental Sciences in general, and students from the Spatial Information Science Laboratory at the University of Tsukuba in particular who have taken this lecture for years. I am truly indebted to the students because I learnt a lot about machine learning and R during the lectures.

I am also indebted to my colleague, Ms. Emiko Ariyasu from Asia Air Survey Co., Ltd. who read the first draft of the workbook. Her comments provided invaluable insights during the revision. I am also truly grateful to the Global Business Development Division management at Asia Air Survey Co., Ltd. who gave me permission to teach part-time at the University of Tsukuba. I extend my special thanks to emeritus professor Masamu Aniya, my academic advisor who provided critical support during my graduate and post-graduate days at the University of Tsukuba. At Springer (Japan), I would like to thank Yosuka Nishida, Asami Komada, and Arun Kumar, and anonymous reviewers.

Last but not least, this workbook would not have been possible without the continued support of my wife Yuko, and my girls Ai and Reina.

Contents

1 Remote Sensing Digital Image Processing in R 1
1.1 Introduction 1
1.1.1 Remote Sensing Digital Image Processing 1
1.1.2 Overview of Machine Learning 3
1.2 Overview of R 6
1.2.1 What Is R? 6
1.2.2 Installing R 6
1.2.3 RStudio 8
1.2.4 R Packages 9
1.2.5 Overview of Data Type, Structure, and Functions 10
1.2.6 Getting Help 11
1.2.7 Handling Errors and Warnings 12
1.2.8 Other Issues 13
1.3 Data and Test Site 13
1.3.1 Landsat Imagery 13
1.3.2 Reference Data Sets 14
1.3.3 Overview of Harare Metropolitan Province 15
1.4 Tutorial 1: A Quick Guide to R 16
1.4.1 Objectives 16
1.4.2 Overview of the Packages 16
1.4.3 R Basics 17
1.4.4 Procedure 17
1.5 Summary 20

1.6 Additional Exercises 21
References 21

2 Pre-processing 25
2.1 Background 25
2.1.1 Radiometric Correction 26
2.2 Tutorial 1: Display Landsat 5 TM Imagery 28
2.2.1 Objectives 28
2.2.2 Overview of Packages 28
2.2.3 Procedure 29
2.3 Tutorial 2: Radiometric Correction and Reprojection 42
2.3.1 Objectives 42
2.3.2 Procedure 42
2.4 Summary 65
2.5 Additional Exercises 65
References 65

3 Image Transformation 67
3.1 Background 67
3.1.1 Spectral Indices 68
3.1.2 Texture Indices 68
3.2 Tutorial 1: Vegetation Indices 69
3.2.1 Objective 69
3.2.2 Overview of the Packages 70
3.2.3 Procedure 70
3.3 Tutorial 2: Texture Analysis 72
3.3.1 Objective 72
3.3.2 Overview of the Packages 73
3.3.3 Procedure 73
3.4 Summary 78
3.5 Additional Exercises 78
References 79

4 Image Classification 81
4.1 Overview of Image Classification 81
4.1.1 k-Nearest Neighbors (KNN) 82
4.1.2 Artificial Neural Networks (ANN) 83

4.1.3 Single Decision Trees (DT) 83
4.1.4 Support Vector Machines (SVM) 84
4.1.5 Random Forest (RF) 85
4.2 Tutorial 1: Single Date Image Classification 85
4.2.1 Objectives 85
4.2.2 Overview of Packages 86
4.2.3 Procedure 87
4.2.4 Summary for Tutorial 1 133
4.3 Tutorial 2: Multidate Landsat Image Classification 134
4.3.1 Objective 134
4.3.2 Procedure 135
4.3.3 Summary for Tutorial 2 150
4.4 Summary 150
4.5 Additional Exercises 151
References 151

5 Improving Image Classification 155
5.1 Overview 156
5.2 Feature Selection 156
5.2.1 Importance of Feature Selection 156
5.2.2 Recursive Feature Elimination (RFE) 157
5.3 Tutorial 1: Image Classification Using Multiple Data Sets 157
5.3.1 Objective 157
5.3.2 Procedure 157
5.3.3 Summary for Tutorial 1 170
5.4 Tutorial 2: Image Classification Using Multiple Data Sets with Feature Selection 170
5.4.1 Objective 170
5.4.2 Procedure 170
5.4.3 Summary for Tutorial 2 180
5.5 Summary 180
5.6 Additional Exercises 181
References 181

Appendix 183

Index 185

Abbreviations and Acronyms

ANN	Artificial Neural Network
ART	Adaptive Resonance Theory
ASM	Angular Second Moment
COST	Cosine Estimation of Atmospheric Transmittance
CRAN	Comprehensive R Archive Network
DEM	Digital Elevation Model
DOS	Dark Object Subtraction
DN	Digital Number
DSG	Department of the Surveyor-General
DT	Decision Trees
EMR	Electromagnetic Radiation
FOSS	Free and Open Source Software
GDAL	Geospatial Data Abstraction Library
GIS	Geographic Information Systems
GLCM	Grey Level Co-occurrence Matrix
GUI	Graphical User Interface
IDE	Integrated Development Environment
KNN	k-Nearest Neighbors
MIR	Mid Infrared
MODTRAN	Moderate Resolution Atmospheric Radiance and Transmittance
ML	Machine Learning
MPL	Multi-Layer Perceptron
MSAVI	Modified Soil Adjusted Vegetation Index
NDVI	Normalized Difference Vegetation Index
NIR	Near Infrared

RF	Random Forests
RFE	Recursive Feature Elimination
SAVI	Soil Adjusted Vegetation Index
SDOS	Simple Dark Object Subtraction
SOM	Self-Organizing Feature Maps
SR	Spatial Resolution
SVM	Support Vector Machines
TIR	Thermal Infrared
TM	Thematic Mapper
UAV	Unmanned Aerial Vehicles
USGS	United States Geological Survey

Chapter 1
Remote Sensing Digital Image Processing in R

Abstract Remote sensing digital image processing and classification provide critical land use/cover and land use/cover change information at multiple spatial and temporal scales. Over the past decades, a plethora of image processing and classification methods have been developed and applied. The purpose of this chapter is to introduce remote sensing digital image processing and machine learning in R. The chapter will cover remote sensing image processing and classification, a brief overview on R and RStudio, tutorial exercises, data and test site.

Keywords Remote sensing · Digital image processing · Machine learning · R · RStudio

1.1 Introduction

1.1.1 Remote Sensing Digital Image Processing

Earth observing remote sensing is the science and art of acquiring information about the Earth's surface using noncontact sensor systems. Sensors on board satellites, aircrafts or unmanned aerial vehicles (UAV) record reflected or emitted electromagnetic radiation (EMR) from features on the Earth's surface. The reflected or emitted electromagnetic radiation (EMR) from features on the Earth's surface is captured as a pixel, which is then converted to a digital number (DN) in order to produce remotely-sensed image. Computers and

C. Kamusoko, *Remote Sensing Image Classification in R*, Springer Geography,
https://doi.org/10.1007/978-981-13-8012-9_1

software are required for digital image processing since the remotely-sensed images are digital. Digital image processing procedures such as pre-processing, image transformation, and image classification will be covered in this workbook. While there many commercial, and free and open source software systems (FOSS) for remote sensing image processing and analysis, R will be used in this workbook.

Remote sensing digital image processing and analysis provide valuable information on land use/cover and its changes, agriculture, environment and other applications at multiple spatial and temporal scales. A variety of methods have been developed and used for land use/cover classification. These classification methods include: the incorporation of structural and textural information (Gong and Howarth 1990; Moller-Jensen 1990); combining satellite images with ancillary data (Harris and Ventura 1995); vegetation—impervious surface—soil models (Ridd 1995); expert systems (Stefanov et al. 2001); hybrid methods that incorporates soft and hard classifications (Lo and Choi 2004); the use of normalized difference built-up index (Xu 2007; Zha et al. 2003); neural networks and deep learning (Seto and Liu 2003; Yu et al. 2017); segmentation and object-based classifications (Guindon et al. 2004); support vector machines (Nemmour and Chibani 2006; Pal and Mather 2003) and random forests (Rodriguez-Galiano et al. 2012).

To date, many advanced classification methods (or classifiers)—in particular machine learning methods—have been used to improve remotely-sensed image classification. However, there are still major challenges such uncertainties in remotely-sensed data, lack of reliable or insufficient reference data sets, and generally a "blind" trust in advanced classification methods (Lu and Weng 2007). Although remote sensing literature is replete with successful applications (Lu and Weng 2007), land use/cover classifications based on medium resolution satellite imagery such as Landsat data (as we will see in our study area) pose many methodological challenges (Griffiths et al. 2010). This is because the urban study area is characterized by a complex and contrasting spatial and socioeconomic development patterns. For example, similar spectral responses between built-up areas on one hand, and bare vacant plots and agriculture areas on the other hand have been observed to cause classification errors.

Furthermore, forest cover classification—which is major input in forestry and climate change projects—is riddled with many uncertainties. For example, sparse forest or woodland areas exhibit high degrees of spatial heterogeneity, which is influenced by soil type, land use, and seasonal changes (Sedano et al. 2005). As a result, remotely-sensed image classification should not be limited to a mere technical approach based on "fancy" or latest machine learning methods. Rather, a more integrated approach that highlights uncertainties in data acquisition (satellite imagery and training data), pre-processing, image classification and accuracy assessment is required. Recent developments such as the use of dense satellite imagery due to the availability of more computer processing power or cloud computing, and more awareness on the need to improve accuracy assessment are positive.

In this workbook, machine learning methods or classifiers such as k-nearest neighbors (KNN), single decision trees (DT), artificial neural networks (ANN), support vector machines (SVM) and random forest (RF) will be used to classify Landsat 5 TM imagery. I have selected these machine learning classfiers since most of them have been widely used for remotely-sensed image classification over the past decades. Note that the idea is not to find the "best method" but to provide hands-on approach to remotely-sensed image processing based on remote sensing and machine learning principles. An attempt is made to focus on common problems related to pre-processing, analysis of data sets prior to image processing and classification, machine learning model tuning and performance (cross-validation), and accuracy assessment.

1.1.2 Overview of Machine Learning

During the past decades, the knowledge domain of machine learning has grown rapidly given the developments in computer technology coupled with advances in information science, statistics and data mining techniques. In addition, the availability of more data from the internet, mobile technology and social media has also contributed to the advancement of machine learning. To date, there are many machine learning classifiers, and applications in diverse fields such as banking and finance, health and medicine, image processing, and environment. While there are many definitions of machine learning, in this workbook we will simply define machine learning as computational methods that learn (or improve their performance) from data (Hastie et al. 2009; Dietterich 1999). According to Mitchell (1997), "a computer program can learn from experience *E* with respect to some class of tasks *T*, and performance measure *P*". Therefore, machine learning algorithms use experience in order to improve performance. Experience refers to data-driven tasks, which are based on statistics and probability (Mitchell 1997).

Machine learning combines statistics, computer science and data mining to solve classification, clustering, regression and other pattern recognition problems (Michie et al. 1994; Ripley 1996; Hastie et al. 2009, Cracknell and Reading 2014). It should be noted from the onset that computer-based statistical approaches are a key component of machine learning (Hastie et al. 2009). However, machine learning approaches differ from conventional statistical approaches. This is because statistical approaches first assume an appropriate data model for fitting, and then model parameters are estimated from the data. In other words, it assumed that the basic form of the model equation and error function is known and therefore, the goal is to find the coefficients of variables in a known function (Hastie et al. 2009). In this regard, the emphasis is on understanding the underlying statistical model and mechanisms as well as model assumptions (Clark 2013). In contrast, machine learning uses an algorithm to learn the relationship between the response (target or dependent) variable and its predictor (independent) variables, particularly in complex data sets (Breiman 2001). As such, the focus of machine learning is more on performance rather than understanding the underlying statistical model mechanisms and assumptions (Hastie et al. 2009; Clark 2013).

Machine learning performs inductive data inference or learning (MacKay 2003; Gahegan 2000), which simply refers to gaining information, knowledge and wisdom from the analysis of raw data (Fotheringham et al. 2002). Inductive inference or learning uses available data or observations to identify patterns. That is, generalizations are derived from specific training data (Bousquet et al. 2004). Essentially, the results obtained from the inference or learning analyses are then applied to other similar data in order to predict, interpret and inform the decision making process (Bousquet et al. 2004). This is in contrast to deductive inference, where a hypothesis (or model) regarding some natural phenomenon is formulated (Gubbins 2004). The observed data are then used to accept or reject the hypothesis.

Machine learning techniques can be classified according to many different criteria, which depend on the nature of the problem that needs to be solved (Russell and Norvig 2003). In this workbook, we will classify the machine learning techniques according to the task or learning style, and the desired output. In terms of task or learning style, machine learning techniques can be grouped into supervised, unsupervised, semi-supervised and reinforcement learning. For supervised learning, example input and output data known as training data are provided. The goal is to predict the outcome based on the input data (Hastie et al. 2009). Training is guided by the minimization of some error based on the internal structure of the learning algorithm (Bousquet et al. 2004; Hastie et al. 2009). For unsupervised learning, the algorithm finds natural groups in data without any labels (Ripley 1996; Hastie et al. 2009). Therefore, the algorithm discovers patterns in data based on clustering. Semi-supervised learning incorporates incomplete training data. That is, both labeled and unlabeled training data are used (Han et al. 2012). The training data are usually composed of a small labeled data set and a large unlabeled data set. Generally, the labeled training data are used to learn, while the unlabeled training data is used to refine the class boundaries (Han et al. 2012). In reinforcement learning, the algorithm interacts in a dynamic environment, whereby learning is based on external positive and negative feedbacks (Portugal et al. 2018). Through trial and error, the algorithm learns to reward positive feedbacks, and avoid negative feedbacks (Portugal et al. 2018).

According to the desired output, machine learning can be categorized as classification, regression, clustering, and density estimation. In classification, inputs are divided into two or more classes, and the learner must produce a model that assigns unseen inputs to one or more of these classes, while in regression the outputs are continuous rather than discrete. Both classification and regression can be supervised. Clustering is an unsupervised task where a set of unknown or unlabeled inputs are divided into groups, while density estimation finds the distribution of inputs in some space.

1.2 Overview of R

1.2.1 What Is R?

R is a free and open source integrated software platform for programming, data manipulation, computation and graphical display (R Development Core Team 2005). According to Maindonald and Braun (2003), R is a dialect of the S language (that was developed at AT&T Bell Laboratories by Rick Becker, John Chambers and Allan Wilks). This script-based software with command lines was developed by Ross Ihaka and Robert Gentleman at the University of Auckland, New Zealand. R was released under the GNU GPL license in 1995, and it has since evolved. Currently, R is now being maintained by a group of developers, and is available on multiple computing platforms such as Windows, GNU/Linux, and Mac OS X. All the R scripts and commands used in this workbook were executed on the Windows platform. However, R scripts and commands can also be adapted for other platforms.

1.2.2 Installing R

In order to install R, go to the R project download link at https://www.r-project.org/ (Fig. 1.1) and click on CRAN (Comprehensive R Archive Network). This website will direct you to a list of international mirror sites from which you can download R (Fig. 1.2). Note that the CRAN mirror sites are organized by country. Therefore it is better to use the mirror site that is closer geographically in order to download R faster. For example, my mirror site is "The Institute of Statistical Mathematics" in Tokyo (https://cran.ism.ac.jp/) because I live near Tokyo. Select a mirror and click the link. A list of R downloads links for Linux, Mac (OS X) and Windows are available (Fig. 1.3). For Windows, you can select **base**, which has binaries for the base distribution if you are installing R for the first time. After that, just double-click the installer "Download R 3.4.3 for Windows (62 megabytes, 32/64 bit)" and follow the instructions. It is easy to install R on Windows, if you have administrative permission. On Windows, R is installed as a graphical user interface

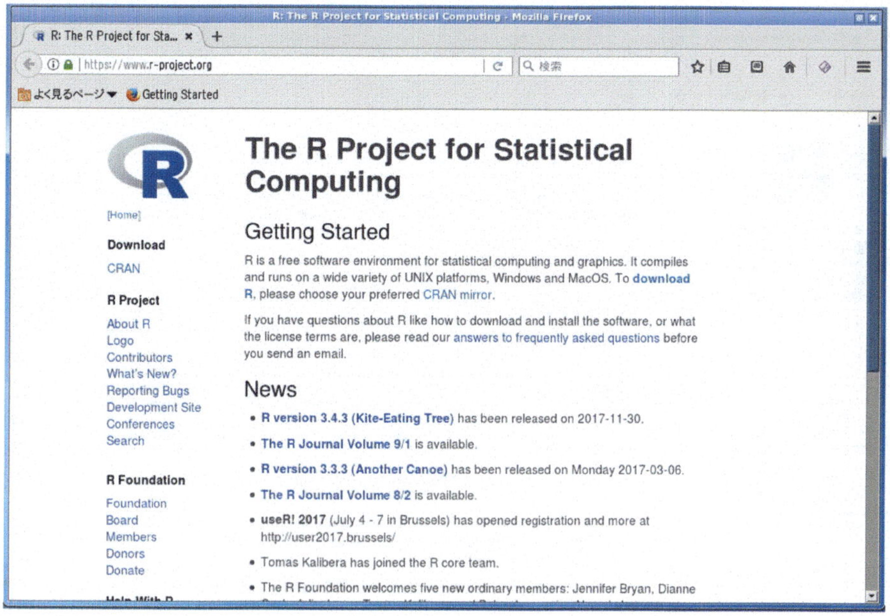

Fig. 1.1 Screenshot of the R Project website

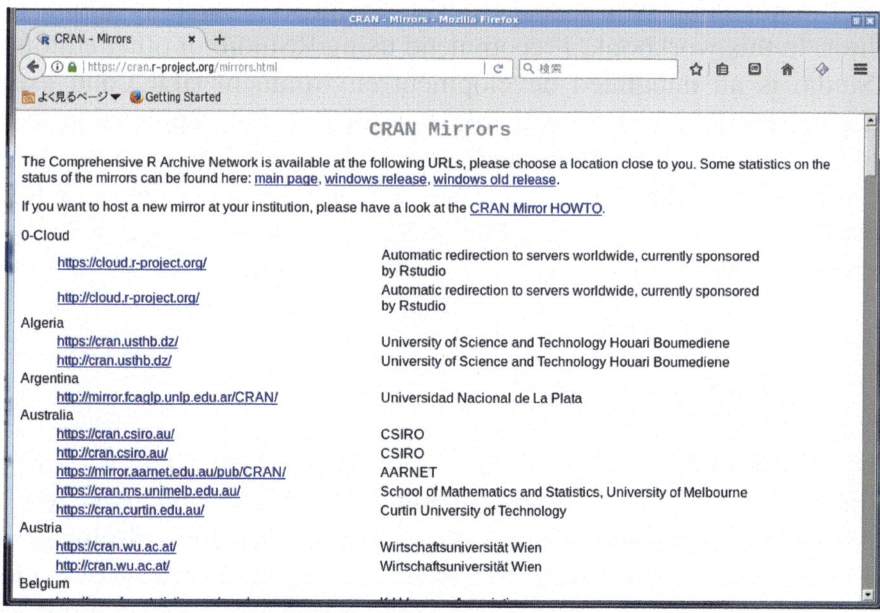

Fig. 1.2 Screenshot of the R CRAN mirrors

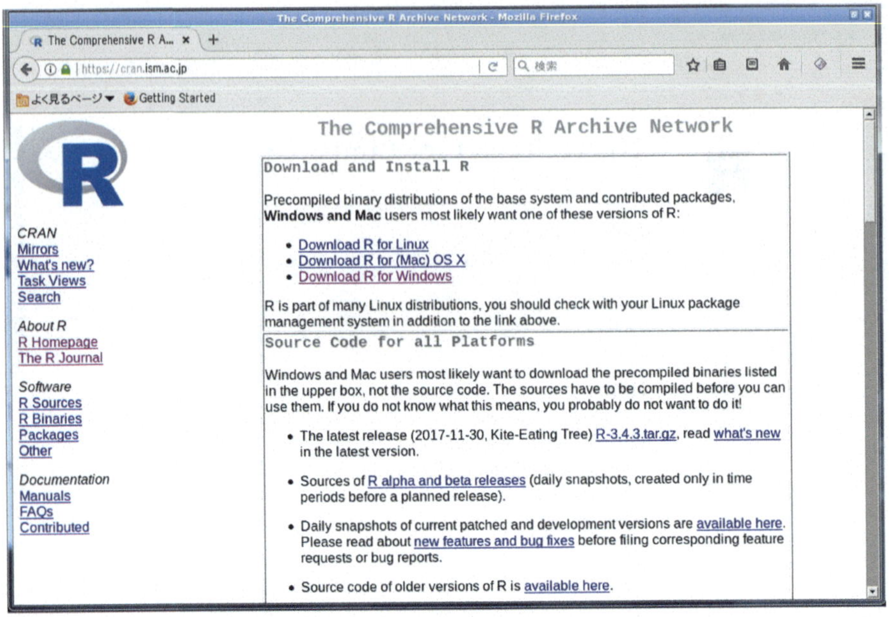

Fig. 1.3 Screenshot of the R CRAN download and install

(GUI) application with a set of development tools such as a built-in editor. In this workbook, I recommend using RStudio or other editors. RStudio is an integrated development environment (IDE) and user interface for R, which is available as a GUI desktop application that runs R on a local machine, and server. You can download the free version of the RStudio at https://www.rstudio.com/products/rstudio/download/. In this workbook, I am using R version 3.4.3 and RStudio version 1.1.383.

1.2.3 RStudio

The RStudio GUI (Fig. 1.4) consists of the Console window, Script Editor Window, and several additional utility windows (e.g., Environment, History, Connections, Plots, Packages, Help, and Viewer). This allows users to easily navigate between views and tools. Note that the script you type in the console window will not be saved. Therefore, it is better to type the commands in the Script Editor

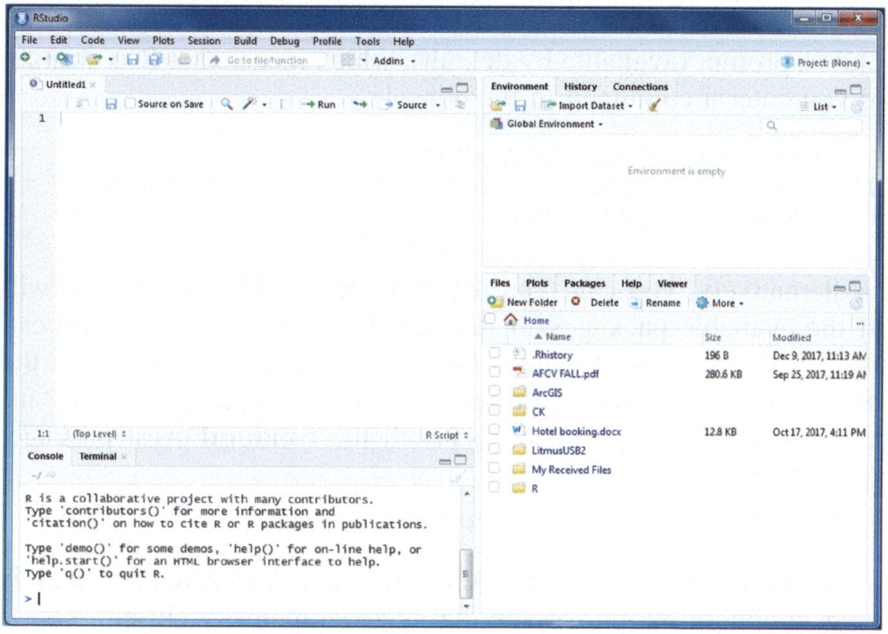

Fig. 1.4 RStudio graphical user interface

in order to save the script file and then reload when you need it in future. More details about RStudio can be found on the RStudio website (https://www.rstudio.com/).

1.2.4 R Packages

R has many packages, which are a set of related functions, help and data files for graphical, statistical and machine learning analysis. The **base** package contains basic arithmetic and statistical functions as well as input and output operations. Note that the **base** package is loaded by default when R is launched. However, you can download other packages from the CRAN mirror (https://cran.r-project.org/) website or other public repositories such as Bioconductor (http://www.bioconductor.org/packages/release/software.html) and install it in your local repository. In order to use a package, you need to check if it is installed into the local library on your computer. You can check the availability of the packages by typing the command below. Note that you do not type the greater than sign ">" in the console. This sign is just a prompt (available by default), which means that R is waiting for you enter a command in the console,

```
> (.packages(all.available=TRUE))
[1]  "kernlab" "randomForest"  "raster"  "RColorBrewer"
  "reshape2"  "rgdal""rgeos"
```

Alternatively, you can type the "***library()***" command, which will list the available packages. If the package is not available, you can install it through the Windows GUI or typing the commands in the console or script editor. I prefer installing the packages using "install.packages". For example, to install the **randomForest** package, type:

```
> install.packages("randomForest")
```

R will ask you to select a CRAN mirror site. As I stated before, it is better to choose the mirror site that is geographically close.

After installing, you can type either the "***library()***" or "***require()***" commands in order to check if package has been correctly installed or is running well.

```
> library(random Forest)

> require(randomForest)
```

If in any case you need to remove the package, you can use the "remove.packages" command as shown below.

```
> remove.packages("randomForest")
```

1.2.5 Overview of Data Type, Structure, and Functions

Generally, variables store data or information in programs such as R. The variables are reserved memory locations that store values in different data types such as character, numerical (integer or floating), boolean (true or false), complex etc. The numeric data type is used to represent floating point numbers while integer data type represents only integer values. In R, variables are assigned with objects such as vectors, arrays, matrices, factors, data frames and lists.

A vector is an ordered row of cells. The concatenate ***c()*** function is used to combine the elements into a vector. An array is multidimensional vector, while a matrix is a two-dimensional rectangular data set or array (Adler 2010). The matrix can be created using a vector input to the matrix function. Factors are objects that store the vector along with the distinct values of the elements in the vector as labels. The labels are always character and are useful in statistical modeling. Data frames are tabular data objects, which have column that can contain different modes of data. For example, the first column can be numeric, the second column can be character and third column can be logical. A list is an R-object which can contain many different types of elements inside it such as vectors, and functions.

R uses "commands" that are comprised of "objects" and "functions", which are typed in the Script Editor or Console as: *object* < - *function*. This simply means that the "object" (which could be a variable or model) is created from a function (that is the operations that do the work). R is an object-oriented language, where each object has a type and class. The type and class defines how the object is stored as well as it operations (that are known as methods). In R functions are provided by the **base** and other packages, or can be user-defined. The basic function is defined as follows:

Name.of.function <- function(argument1, argument2) {
tatements
Return(something)}

Furthermore, R provides arithmetic operators such as addition (+), subtraction (−), multiplication (*), division (/), and exponentiation (^). Note that detailed explanations of objects, methods, classes and functions are beyond the scope of this workbook. Therefore, readers are encouraged to read the R manual (R Development Core Team 2005) and other introductory books such as Adler (2010).

1.2.6 Getting Help

R has a good help system that you can use to get information about the installed packages and functions. For example, to get help on the **RStoolbox** package, you can simply type the following commands:

```
> ??RStoolbox
> help(RStoolbox)
```

In R, functions have arguments, which are specific slots that are used to pass objects (Adler 2010; Kuhn and Johnson 2016). Note that in most cases, I use default function settings or arguments. However, readers can access the help system in order to learn about additional arguments or function options. For example, you can simply type "?? rfe" in the console if you want to know about the functions in the recursive feature elimination (RFE) model.

1.2.7 Handling Errors and Warnings

In most cases, first time users of R will encounter errors and warnings as they run commands and scripts. Generally, errors stop the execution of the commands, while warnings do not. Although a warning allows the command to be executed, the warning message or mesages should be read carefully in order to avoid mistakes or false results.

The first time I used R, I encountered many file path, syntax and missing object errors. The file path errors are related to specifying a wrong file path. Examples of syntax errors are missing commas, unmatched parentheses, and the wrong type of closing brace. The missing object errors are related to wrong spellings or capitalization (note that R is case sensitive). Sometimes, the file containing the object is not on the search list or in some cases the required packages are not simply installed or loaded. Furthermore, R is updated regularly. As a result, some packages may not be compatible with your current R software version. This may also cause errors.

Note that errors or problems will occur most of the times (especially for beginners). So do not panick. Remember, the best way to learn is through errors or mistakes. However, it is important to read the error messages carefully, and then attempt to solve the problem. You can also check for solutions on the R website or other online sources for help and tips (see appendix). I usually consult stackoverflow for tips and advice whenever I encounter problems.

1.2.8 Other Issues

Finally, in R data is read or imported using commands that are kept in memory. However, there is a choice to save the workspace to disk or select objects that can be saved so that they can be loaded in the next session. Note that saving the workspace is good practice for documenting work. Last but not least, there many ways to perform analysis in R. Therefore, readers or analysts can try alternative ways or better methods in order to improve data analysis in R.

1.3 Data and Test Site

Landsat 5 Thematic Mapper (TM) imagery acquired in 1984 (Table 1.1) will be used for all image processing and classification. The Landsat 5 TM imagery was downloaded from the United States Geological Survey (USGS) website. Harare Metropolitan Province in Zimbabwe will be used as a test site in this book. Following is a brief description of the Landsat imagery, reference data sets and test site.

1.3.1 Landsat Imagery

In this workbook, Landsat 5 TM imagery was acquired for two dates (hereafter referred to as multidate Landsat 5 TM imagery) corresponding to the dry and early wet seasons (Table 1.1 and Fig. 1.5). The multidate Landsat 5 TM imagery was used for image processing and classification since it accounts for seasonality or vegetation phenology (Fig. 1.5). Table 1.2 shows the characteristics of the Landsat 5 TM sensor.

Table 1.1 Summary of Landsat 5 TM imagery used

Sensor	Date	Path/row	Season
Landsat 5 TM	22-06-1984	170/072	Dry
Landsat 5 TM	31-12-1984	170/072	Wet

Fig. 1.5 Landsat 5 TM imagery for the test site. Note that, the Landsat 5 TM imagery on the left side was acquired in the dry season, while the one on the right side was acquired during the wet season

Table 1.2 Characteristics of Landsat 5 TM sensor

Band	Wavelength (μm)	Electromagnetic spectrum	SR (m)	Applications
1	0.45–0.52	Blue (B)	30	Coastal water mapping
2	0.52–0.60	Green (G)	30	Vegetation vigor
3	0.63–0.69	Red (R)	30	Chlorophyll absorption
4	0.76–0.90	Near-infrared (NIR)	30	Vegetation types
5	1.55–1.75	Mid-infrared (MIR)	30	Moisture/soils
6	10.40–12.5	Thermal infrared (TIR)	120	Thermal mapping
7	2.08–2.35	Mid-infrared (MIR)	30	Minerals/soils

Note All TM bands are quantized as 8 bit data (values for 0–255). SR refers to spatial resolution

1.3.2 Reference Data Sets

Reference data sets that consist of classifier training and classification accuracy assessment points were generated from black and white aerial photographs (at a scale of 1:25,000), which were acquired in 1984. These black and white aerial photographs were obtained from the Department of the Surveyor-General (DSG), Zimbabwe. The reference data sets were originally compiled for a pilot urban land use/cover project (Kamusoko et al. 2013), which focused on the classification of built-up and non-built up areas. Therefore, there is more reference data on settlement, bareland and agriculture areas than green space and water areas. A lot effort was made to prepare reliable and accurate reference data sets. However, it is important to recognize that reference data sets were generated from black and white aerial

Table 1.3 Land use/cover classification scheme

Land use/cover	Description
Settlement	Residential, commercial and services, industrial, transportation, communication and utilities and construction sites
Green space	Woodlands, riverine vegetation, shrub and bush, grass cover, open grass areas, golf courses, and parks
Agriculture	Cultivated land or land being prepared for raising crops, fallow agriculture land, and land under irrigation
Bareland	Bare exposed areas and transitional areas, and excavation sites
Water	Rivers and reservoirs

photographs acquired in the dry season. In addition, the black and white aerial photographs were scanned and then georeferenced. Therefore, it is inevitable that some errors are found within the reference data set. Nonetheless, the reference data set is very useful since it was derived from near-anniversary aerial photographs that were acquired in the dry season (same as the dry season Landsat 5 TM imagery). Table 1.3 shows the adopted land use/cover classes, which are based on the "Forestry Commission (Zimbabwe) and DSG national woody cover classes" classification schemes, and the author's a priori knowledge of the test site.

1.3.3 Overview of Harare Metropolitan Province[1]

Harare metropolitan province comprises four districts namely, Harare Urban, Harare Rural, Chitungwiza and Epworth. The metropolitan province covers about 942 km^2. The average altitude is approximately 1,500 m above sea level. The study area is characterized by a warm, wet season from November to April; a cool, dry season from May to August; and a hot, dry season in October. Daily temperatures range from about 7–20 °C in July (coldest month), and from 13 to 28 °C in October (hottest month). The test site receives a mean annual rainfall ranging from 470 to 1,350 mm between November and March. Vegetation varies from grasslands to open Miombo

[1]The "Overview of Harare Metropolitan Province" section is modified from Kamusoko et al. 2013,

woodlands dominated by Brachystegia spiciformis trees as well as some introduced tree species such as Jacaranda. The metropolitan province is dominated by a complex of: gabbro and dolerite to the north; an intrusion of metagreywacke and phylite in the center; and granites to the east, and southwest. The soils in the test site are dominated by fersialitic and paraferrallitic soils (Nyamapfene 1991).

1.4 Tutorial 1: A Quick Guide to R

1.4.1 Objectives

- Launch R, and
- Interact with simple expressions and functions in R Console environment.

1.4.2 Overview of the Packages

As I mentioned in Sect. 1.2.4, R comes with packages that are loaded by default such as **base**, **graphics**, and **stats**. The first tutorial exercise

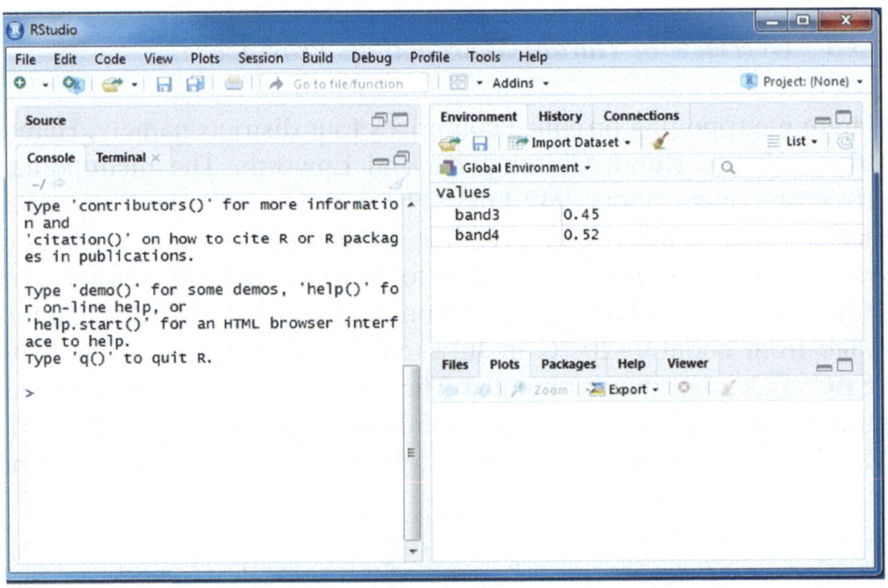

Fig. 1.6 RStudio interface

will use the **base** and **stats** packages. The **base** package contains basic functions in R that include arithmetic, while the **stats** package includes functions for computing statistics and some modeling tools.

1.4.3 R Basics

Before we start working, let's have a brief introduction on R basics. In R, you can assign a value to an object using "<-" or "=". In order to avoid confusion, I will use the assignment operator "<-" to assign values to variables or creating objects. The sign "=" will be used to specify values to function arguments in this workbook. For example, *classified <- gbm(object, type = "prob").*

1.4.4 Procedure

Step 1: Launch RStudio

Navigate to **Start-Program** and then double-click **RStudio** to launch the program (Fig. 1.6). We will start using RStudio in interactive mode (that is, in R Console).

Step 2: Working with operators and simple expressions in R

Figure 1.6 shows the R Console screenshot. You can see the prompt (>), which indicates that R is waiting for the command. Let's start by typing the expression 2 + 3 in the R Console (Fig. 1.7).

```
> 2 + 3
```

The result '5' is returned, when we execute the above command (Fig. 1.7).

```
[1] 5
```

Next, let's create an expression using variables. We need to write the following expressions: "band3 <- 0.45" and "band4 <- 0.52". This simply means that I am assigning the reflectance value of 0.45 to band3 (Red), and reflectance value of 0.52 to band4 (NIR).

First, assign the reflectance value 0.45 to band3, and 0.52 to band4. Then press enter.

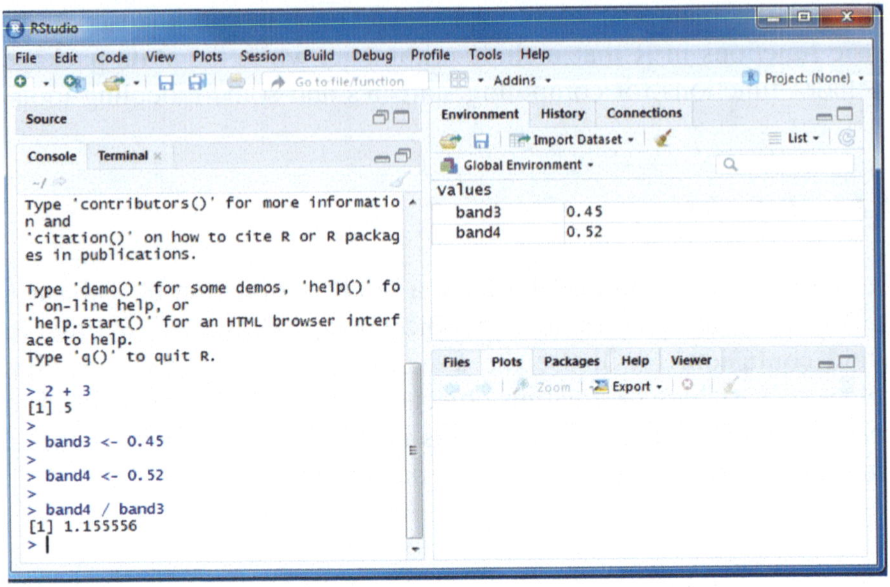

Fig. 1.7 Basic expressions in RStudio interface

```
> band3 <- 0.45
> band4 <- 0.52
```

Finally, type "band4 / band3" using the division (/) operator and press enter.

```
> band4 / band3
```

The following output is produced when we execute the above commands (Fig. 1.7).

```
[1] 1.155556
```

Note that in R, you can use the dot character (.) or underscore (_) for naming variables. For example, both band.4 and band_4 can be used.

Step 3: Create a vector.

The concatenate function, ***c()*** is used to combine the elements into a vector as shown in the code below.

```
> bands <- c("red","green","blue")
> print(bands)
```

The following output is produced when the commands are executed.

```
[1] "red"   "green" "blue"
```

Step 4: Create a matrix

We can use the ***matrix()*** and ***c()*** functions to create a matrix that contains band reflectance values as shown below.

```
> band_ref = matrix( c(0.32,0.45,0.55,0.22,.062,0.82),
nrow = 2, ncol = 3, byrow = T)
> print(band_ref)
```

The following output is produced when the commands are executed.

```
     [,1]  [,2] [,3]
[1,] 0.32 0.450 0.55
[2,] 0.22 0.062 0.82
```

Step 5: Create a data frame

Let's create a data frame that contains reflectance values (reflvalues) using ***data.frame()***.

```
> reflValues <- data.frame(Land_cover = c("Agriculture",
 "Forest","Bareland","Built-up","Water"),
        Band1 = c(0.62, 0.41, 0.71, 0.52, 0.11),
        Band2 = c(0.52, 0.61, 0.21, 0.32, 0.41),
        Band3= c(0.82, 0.21, 0.61, 0.42, 0.71),
        Band4 = c(0.42, 0.61, 0.31, 0.12, 0.21),
        Band5 = c(0.02, 0.31, 0.21, 0.82, 0.31),
        Band7= c(0.12, 0.21, 0.41, 0.12, 0.51))

> print(reflValues)
```

The following output is produced when the command is executed.

```
   Land_cover Band1 Band2 Band3 Band4 Band5 Band7
1 Agriculture  0.62  0.52  0.82  0.42  0.02  0.12
2      Forest  0.41  0.61  0.21  0.61  0.31  0.21
3    Bareland  0.71  0.21  0.61  0.31  0.21  0.41
4    Built-up  0.52  0.32  0.42  0.12  0.82  0.12
5       Water  0.11  0.41  0.71  0.21  0.31  0.51
```

Step 6: Create functions

As I stated before, in R functions are either provided by **base** package and other packages or can be user-defined. First, let's use the ***mean()*** function from the **base** package to calculate mean spectral reflectance.

```
> refl.values = c(0.27, 0.34, 0.45, 0.53, 0.65)
> mean(refl.values)
```

The following output is produced when the commands are executed.

```
[1] 0.448
```

Next, let's define a simple function to calculate the normalized difference vegetation index (NDVI). In order to do that, we need to declare the variables for band 3 (Red) and band 4 (NIR). The NDVI formula is expressed as: NDVI = NIR – Red/NIR + Red. First, let's assign reflectance values to bands 3 and 4.

```
> band3 = 0.45

> band4 = 0.52
```

We can now define NDVI by assigning it to the output function. The list of argument names are contained within parentheses. Next, the body of the function–the statements that are executed–is contained within curly braces ({ }).

```
> ndvi <- function(band3, band4) {
     band_ratio <- ((band4 -band3) / (band4 + band3))
     return(band_ratio)}
```

When we call the ***ndvi()*** function, the values we pass to it are assigned to those variables so that we can use them inside the function.

```
> ndvi(band3, band4)
```

The following output is produced when we execute the code.

```
[1] 0.07216495
```

1.5 Summary

This chapter provided an overview of remote sensing digital image processing, machine learning, R and RStudio. In addition, the data sets and test site were also briefly described. Last but not least, we managed to run R in the interactive mode. The next chapter will cover remote sensing image pre-processing.

1.6 Additional Exercises

(i) Create a simple function to calculate the following simple band ratios:

- TM3/TM2
- TM4/TM5
- TM5/TM7.

References

Adler J (2010) R in a Nutshell. O'Reilly Media

Bousquet O, Boucheron S, Lugosi G (2004) Introduction to statistical learning theory. In: Bousquet O, von Luxburg U, Ratsch G (eds) Advanced lectures on machine learning. Lecture notes in artificial intelligence, vol 3176. Springer, Heidelberg, Germany

Breiman L (2001) Random forests. Mach Learn 45:5–32

Clarke B (2013) Guest editorial for Special Issue of Statistical Analysis and Data Mining. 6(4):271–272

Cracknell MJ, Reading AM (2014) Geological mapping using remote sensing data: a comparison of five machine learning algorithms, their response to variations in the spatial distribution of training data and the use of explicit spatial information. Comput Geosci 63:22–33

Dietterich TG (1999) Machine learning. In: Wilson R, Keil F (eds) The MIT encyclopedia of the cognitive sciences. MIT Press, pp 497–498

Fotheringham AS, Brunsdon C, Charlton M (2002) Geographically weighted regression—the analysis of spatially varying relationships. Wiley, Chichester

Gahegan M (2000) On the application of inductive machine learning tools to geographical analysis. Geog Anal 32(2):113–139

Gong P, Howarth PJ (1990) The use of structural information for improving land-cover classification accuracies at the rural-urban fringe. Photogramm Eng Remote Sens 56(1):67–73

Griffiths P, Hostert P, Gruebner O, Linden S (2010) Mapping megacity growth with multi-sensor data. Remote Sens Environ 114 (2):426–439

Gubbins D (2004) Time series analysis and inverse theory for geophysicists. Cambridge University Press, Cambridge

Guindon B, Zhang Y, Dillabaugh C (2004) Landsat urban mapping based on a combined spectral-spatial methodology. Remote Sens Environ 92:218–232

Han J, Kamber M, Pei J (2012) Data mining: concepts and techniques, 3rd edn. Elsevier, Amsterdam

Harris PM, Ventura SJ (1995) The integration of geographic data with remotely-sensed imagery to improve classification in urban area. Photogramm. Eng. Remote Sens 61:993–998

Hastie T, Tibshirani R, Friedman JH (2009) The elements of statistical learning: data mining, inference, and prediction. Springer-Verlag, New York

Kamusoko C, Gamba J, Murakami H (2013) Monitoring urban spatial growth in Harare Metropolitan province, Zimbabwe. Adv Remote Sens 2:322–331

Kuhn M, Johnson K (2016) Applied predictive modeling. Springer

Lo CP, Choi J (2004) A hybrid approach to urban land use/cover mapping using Landsat 7 enhanced thematic mapper plus (ETM+) images. Int J Remote Sens 25:2687–2700

Lu D, Weng Q (2007) A survey of image classification methods and techniques for improving classification performance. Int J Remote Sens 28:823–870

MacKay DJC (2003) Information theory, inference, and learning algorithms. Cambridge University Press. http://www.inference.org.uk/itprnn/book.pdf. Accessed 4 Jan 2018

Maindonald J, Braun J (2003) Data analysis and graphics using R—an example-based approach. Cambridge series in statistical and probabilistic mathematics. Cambridge University Press, Cambridge, UK

Michie D, Spiegelhalter DJ, Taylor CC (1994) Machine learning, neural and statistical classification. Ellis Horwood

Mitchell T (1997) Machine learning. McGraw Hill

Moller-Jensen L (1990) Knowledge-based classification of classification of an urban area using texture and context information in

Landsat-TM imagery. Photogramm Eng Remote Sens 56(6):899–904

Nemmour H, Chibani Y (2006) Multiple support vector machines for land cover change detection: an application for mapping urban extensions. ISPRS J Photogramm Remote Sens 61:125–133

Nyamapfene K (1991) Soils of Zimbabwe. Nehanda Publishers, Harare, Zimbabwe

Pal M, Mather PM (2003) An assessment of the effectiveness of decision tree methods for land cover classification. Remote Sens Environ 86(4):554–565

Portugal I, Alencar P, Cowan D (2018) The use of machine learning in recommender systems: a systematic review. Expert Syst Appl 97:205–227

R Development Core Team (2005) R: A language and environment for statistical computing. R foundation for statistical computing. http://r-project.kr/sites/default/files/2%EA%B0%95%EA%B0%95%EC%A2%8C%EC%86%8C%EA%B0%9C_%EC%8B%A0%EC%A2%85%ED%99%94.pdf. Accessed 3 April 2014

Ridd MK (1995) Exploring a V-I-S (vegetation-impervious surface-soil) model for urban ecosystem analysis through remote sensing: comparative anatomy for cities. Int J Remote Sens 16:2165–2185

Ripley BD (1996) Pattern recognition and neural networks. Cambridge University Press

Rodriguez-Galiano VF, Chica-Olmo M, Abarca-Hernandez F, Atkinson PM, Jeganathan C (2012) Random forest classification of mediterranean land cover using multi-seasonal imagery and multi-seasonal texture. Remote Sens Environ 121:93–107

Russell S, Norvig P (2003) Artificial intelligence: a modern approach, 2nd ed. Prentice Hall

Sedano F, Gong P, Ferrao M (2005) Land cover assessment with MODIS imagery in southern African Miombo ecosystems. Remote Sens Environ 98(4):429–441

Seto KC, Liu W (2003) Comparing ARTMAP neural network with the maximum-likelihood classifier for detecting urban change. Photogramm Eng Remote Sens 69(9):981–990

Stefanov WL, Ramsey MS, Christensen PR (2001) Monitoring urban land cover change: an expert system approach to land cover classification of semiarid to arid centers. Remote Sens Environ 77 (2):173–185

Xu H (2007) Extraction of urban built-up land features from Landsat imagery using a thematic-oriented index combination technique. Photogramm Eng Remote Sens 73:1381–1391

Yu X, Wu X, Luo C, Ren P (2017) Deep learning in remote sensing scene classification: a data augmentation enhanced conventional neural network framework. GIS Remote Sens 54(5):741–758

Zha Y, Gao J, Ni S (2003) Use of normalized difference built-index in automatically mapping urban areas from TM imagery. Int J Remote Sens 24(3):583–594

Chapter 2
Pre-processing

Abstract Remotely-sensed images are influenced by distortions due to sensor, solar, atmospheric and topographic effects. As a result, pre-processing is required in order to minimize sensor, solar, atmospheric and topographic effects. Generally, pre-processing focuses on radiometric and geometric correction (in particular georeferencing and image registration) of remotely-sensed imagery prior to image transformation or classification. Radiometric correction focuses on improving the accuracy of surface reflectance, emittance and back scattering measurements. The aim of geometric correction is to minimize geometric distortions due to sensor-Earth geometry variations, and then converting remotely-sensed imagery to real world coordinates on the Earth's surface. In this workbook, pre-processing will focus mainly on radiometric correction and reprojection of Landsat imagery.

Keywords Radiometric correction · Geometric correction · Surface reflectance · Reprojection

2.1 Background

Remotely-sensed images are influenced by distortions due to sensor, solar, atmospheric and topographic effects. Therefore, pre-processing is performed to minimize these effects, depending of course on each remote sensing application or project specifications. In most remote

Electronic supplementary material The online version of this chapter (https://doi.org/10.1007/978-981-13-8012-9_2) contains supplementary material, which is available to authorized users.

C. Kamusoko, *Remote Sensing Image Classification in R*, Springer Geography,
https://doi.org/10.1007/978-981-13-8012-9_2

sensing projects and applications, pre-processing focuses on radiometric and geometric correction (in particular georeferencing and image registration) of remotely-sensed imagery prior to image transformation or classification (Jensen 2005). The purpose of radiometric correction is to improve the accuracy of surface reflectance, emittance and back scattering measurements. That is, the remotely-sensed data is corrected for sensor internal errors and atmospheric noise. Geometric correction is a procedure for minimizing geometric distortions due to sensor-Earth geometry variations. Therefore, geometric correction converts the remotely-sensed imagery to real world coordinates (e.g., latitude and longitude) on the Earth's surface. In this workbook, pre-processing will focus on radiometric correction and reprojection of Landsat imagery. Following is a brief description of radiometric correction procedures.

2.1.1 Radiometric Correction

The atmosphere affects radiance received at the satellite through scattering, absorption, and refraction (Chavez 1996). Radiometric correction is performed in order to remove or minimize atmospheric effects. The decision to perform radiometric correction depends on: (i) the nature of the problem under study; (ii) the type of remotely-sensed imagery available; (iii) the amount of in situ atmospheric data available; and (iv) the accuracy of the biophysical information that must be obtained from the remotely-sensed imagery (Jensen 2005). A number of absolute and relative radiometric correction models are available (Du et al. 2002).

Absolute or physically-based calibration models are the most complex since in situ atmospheric data and radiative transfer models are used. While these models are considered to be more accurate, they depend on in situ atmospheric data, which is not always available (Du et al. 2002). Examples include 6S (Second Simulation of the Satellite Signal in the Solar Spectrum) and MODTRAN (Moderate Resolution Atmospheric Radiance and Transmittance). Relative or image-based calibration models (Chavez 1996) depend on digital image information, which does not require in situ atmospheric data. The relative calibration models are based on the assumption that multiple looks on the same object from different vintage points or multiple bands should cancel out atmospheric effects (Jensen 2005). However, the atmosphere paths from the multiple looks may not be the same. Examples of image-based and relative calibration models include histogram adjustment of single date imagery, and multidate image normalization based on histogram matching or regression modeling (Du et al. 2002; Jensen 2005).

In this chapter, we perform radiometric correction for Landsat 5 TM imagery using apparent reflectance models, simple dark object subtraction (SDOS), dark object subtraction (DOS), and Cosine estimation of atmospheric transmittance (COST) that are available in RStoolbox (Leutner and Horning 2016). Note that a detailed description or literature review on radiometric correction is beyond the scope of this workbook. Therefore, those who want to learn more should consult the relevant textbook and papers listed in the reference section. Below is a brief description of the radiometric correction models.

(i) Apparent reflectance model

The apparent reflectance model is the simplest model used to convert at-satellite radiances to reflectance. While the model corrects for spectral band solar irradiance and solar zenith angle of the image, it does not correct for atmospheric scattering and absorption. Furthermore, the correction is based on the sensor's gain and offset values.

(ii) Simple dark object subtraction (SDOS)

The SDOS model is based on the assumption that dark objects exist within an image and have zero reflectance. That is, the dark objects reflect no light, and therefore any value greater than zero is from atmospheric scattering or proportional to atmosphere path radiance (which is the additive effects of atmospheric scattering). Generally, the dark object subtraction searches each band for the darkest pixel value and scattering is removed by subtracting this value from every pixel in the band.

(iii) Dark object subtraction (DOS) model

The DOS model corrects for the atmospheric additive scattering as well as spectral band irradiance, and solar zenith. For the DOS model, haze is not estimated for each band separately. While DOS does not provide correct absolute reflectance values, this model provides a good balance between simplicity and accuracy (Chavez 1996).

(iv) Cosine estimation of atmospheric transmittance (COST) model

The COST model was developed to account for multiplicative effects of atmospheric scattering and absorption (Chavez 1996). The cosine function of the solar zenith angle or satellite view angle is used to approximate atmospheric transmittance.

2.2 Tutorial 1: Display Landsat 5 TM Imagery

2.2.1 Objectives

- Import and display Landsat 5 TM imagery acquired on 22 June 1984, and
- Explore basic Landsat 5 TM imagery statistics.

2.2.2 Overview of Packages

The first exercise will use the **sp**, **rgdal**, **raster**, **RStoolbox**, **ggplot2**, and **gridExtra** packages. Table 2.1 provides a brief description of the packages that will be used in tutorial exercise 1.

Table 2.1 Key packages used

Package name	Description
sp	This package provides methods and classes for spatial data (sp objects) such as SpatialPoints, SpatialPolygons or SpatialPixels. In addition, the **sp** package provides classes for projection, data management and querrying. The package is automatically loaded by the raster package
rgdal	This package is R's interface to the "Geospatial Data Abstraction Library (GDAL)", which is used by other open source GIS packages to handle spatial data formats. It provides important functions to import and export spatial data in formats such as shapefile and GeoTIFF. This also includes project transformations
raster	This package provides important functions for importing and handling raster datasets. The **raster** package provides: (i) ***raster()*** function for reading single raster layer; (ii) ***brick()*** function for reading multiple layers from a single file such as a multilayer GeoTIFF; (iii) ***stack()*** for reading seperate files (e.g., from metadata)

(continued)

Table 2.1 (continued)

Package name	Description
RStoolbox	This package provides powerful functions for remote sensing data analysis such as satellite imagery pre-processing (e.g., radiometric correction), analysis (spectral indices and tasselled cap analysis) and land use/cover classification (unsupervised and supervised)
ggplot2	This package provides a plotting system that is based on the grammar of graphics, which are useful prior to data analysis
gridExtra	This package provides a number of user-level functions to work with "grid" graphics (that is, to arrange multiple grid-based plots on a page

Notes Raster data can be represented in R as raster (that is, a single layer raster object), stack (that is, a multiple layer raster object), and brick (that is, a multiple layer raster object). While brick is faster, it is less flexible than raster because all layers must be from the same file

2.2.3 Procedure

In Chap. 1, we created objects and functions in R using the interactive mode. Commands were typed in R Console. From chapter 2 onwards, we will be typing commands in the R Script editor in RStudio so that we can load and reuse the scripts again. Note that, we will be using a subset of the Landsat 5 TM scene to allow fast uploading and processing when excuting the commands.

Step 1: Editing and executing commands

First, click **File** and use the **File -> New File** menu in order to create a new file (Fig. 2.1). This is an efficient way of working with R since it much easier to reproduce sequences of commands or functions that can be used again. If you are using an existing or previously saved file, you use either the **File -> Open File**… menu or the **Recent Files** menu to select from recently opened files.

In R, you can run the commands line by line, as multiple lines or the whole document. In order to execute the line of source code where the cursor currently resides you press the **Ctrl + Enter** key or use the **Run** in the toolbar button (Fig. 2.2). After executing the line of code, RStudio automatically advances the cursor to the next line. Generally,

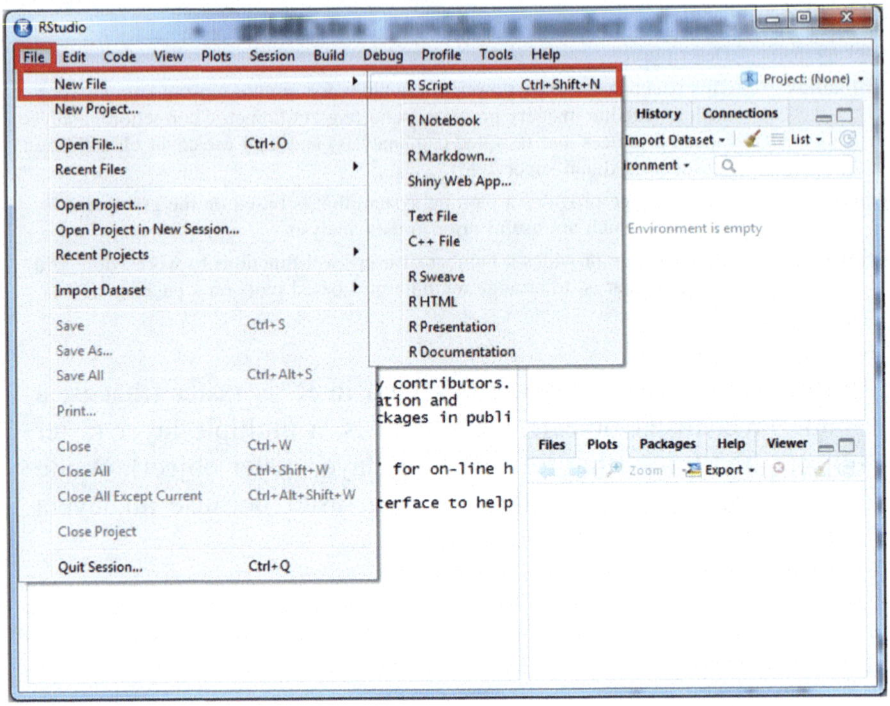

Fig. 2.1 Opening a new script file in RStudio

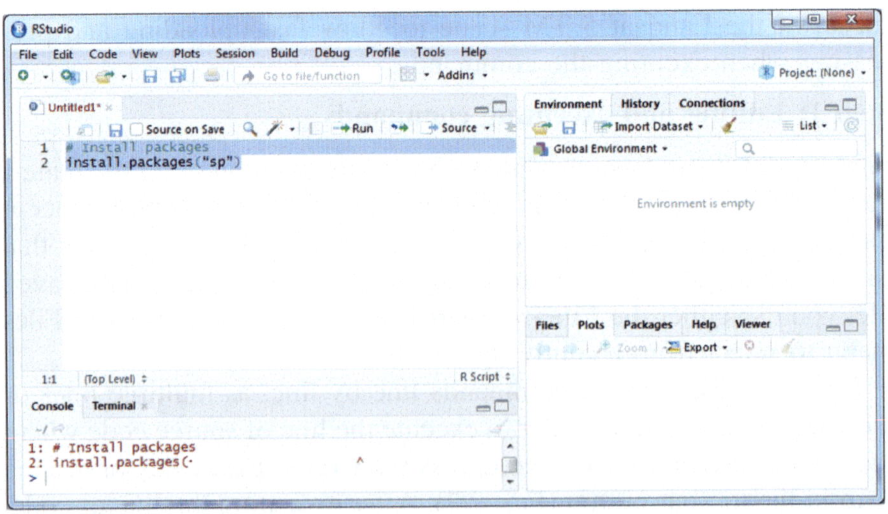

Fig. 2.2 Running the code in RStudio

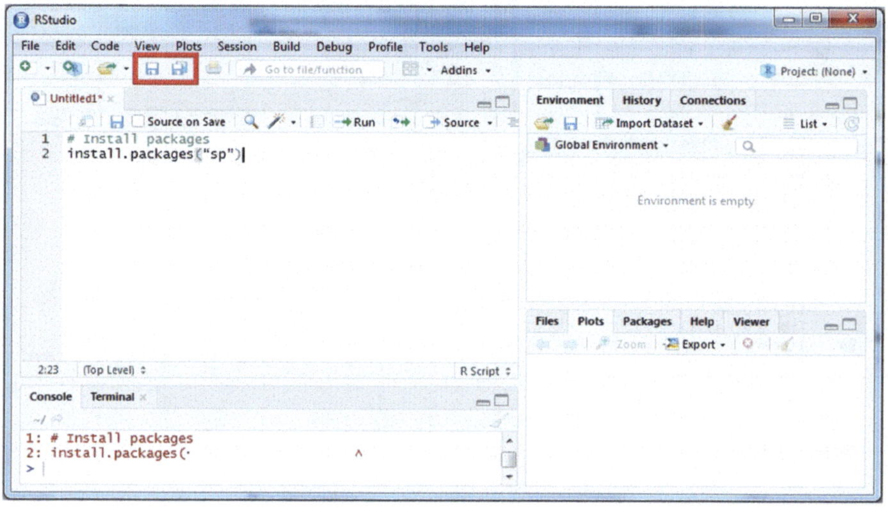

Fig. 2.3 Save the script in RStudio

there are three ways to execute multiple lines from within the editor. However, I usually highlight or select the lines and press the **Ctrl + Enter** key (or use the Run toolbar button). In cases, where I have few lines of code, I run the entire document by pressing the **Ctrl + Shift + Enter** key (or use the Source toolbar button).

Finally, you can save your code using the **Save** or **Save As** button in toolbar shown in Fig. 2.3.

Step 2: Load the necessary packages or libraries

We want to import and display a subset of the Landsat 5 TM imagery, which was on acquired on 22 June 1984. First, you need to install the required packages (Fig. 2.4). Note that some packages may already be installed on your computer. So you should check if the package is installed by loading it using the ***library()*** function as explained in Sect. 1.2. If the packages are not available, you can install them using the following commands:

```
> install.packages("sp")
> install.packages("rgdal")
> install.packages("raster")
> install.packages("RStoolbox")
> install.packages("ggplot2")
> install.packages("gridExtra")
```

Note do not type the arrow sign > in the R Script editor.

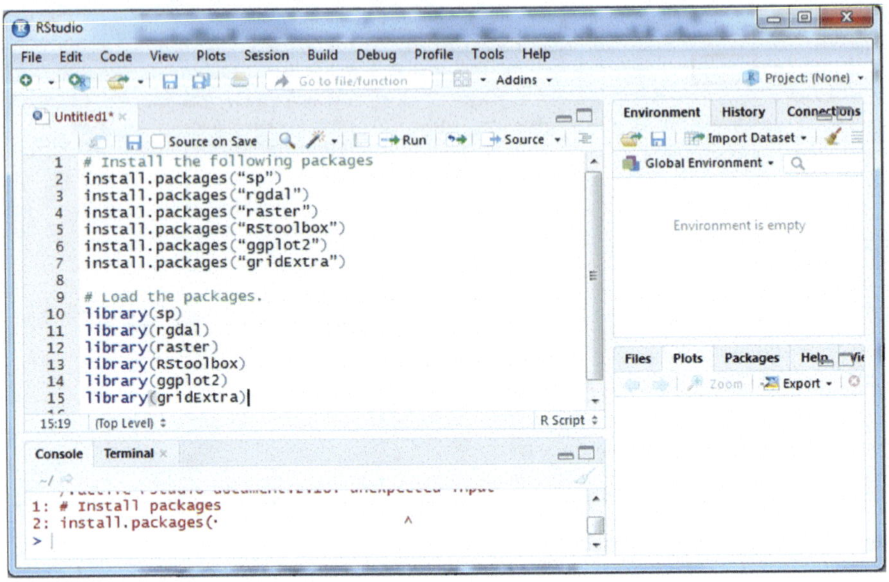

Fig. 2.4 Install and load packages in RScript editor

After installing the packages, load the packages by typing the following commands (Fig. 2.4).

```
> library(sp)
> library(rgdal)
> library(raster)
> library(RStoolbox)
> library(ggplot2)
> library(gridExtra)
```

Step 3: Set up the working directory

Next, setup your working directory using ***setwd()*** command. The working directory is where the downloaded Landsat imagery and other geospatial data sets are stored. For example, my downloaded Landsat 5 TM data is in folder E. Note that the separator in the path names can be either double backslashes or forward slashes. For example:

```
> setwd("E:\\Data\\Original_Subsets")

> setwd("E:/Data/Original_Subsets")
```

In this workbook, I am using the double backslashes.

Step 4: Load Landsat 5 TM scene

We are going to import a subset of the Landsat 5 TM scene. Use the ***readMeta()*** function to import the Landsat 5 TM scene acquired on 22 June 1984. The ***readMeta()*** function reads metadata, and in particular Landsat metadata files.

```
> metaData<- readMeta("LT51700721984174XXX08_MTL.txt")
```

We have created an object "metadata" that comprises Landsat 5 TM bands from "*LT51700721984174XXX08_MTL.txt*" using the ***readMeta()*** function. This function, which is available from the RStoolbox reads metadata. The ***readMeta()*** function accepts two arguments "file" and "raw".

readMeta(*file*; *raw* = *TRUE*)

The "file" is the name or the path to the Landsat MTL file (*LT51700721984174XXX08_MTL.txt*), while "*raw*" is a logical parameter. That is, if TRUE the full raw metadata will be returned as a list. If FALSE (the default) all important metadata are grouped into a standard format. Therefore, you do not type "*raw* = *FALSE*" since R will load all the metadata by default.

Step 5: Check the attributes of the Landsat image files

Run the ***summary()*** function in order to check the attributes of the Landsat metadata. The ***summary()*** function is generally used to display summary results from various model fitting functions.

```
> summary(metaData)
```

The output shows the scene id, type of satellite and sensor, acquisition date, path/row, projection and data are shown. The Landsat 5 TM scene has seven bands.

```
Scene:        LT51700721984174XXX08
Satellite:    LANDSAT5
Sensor:       TM
Date:         1984-06-22
Path/Row:     170/72
Projection: +proj=utm +zone=36 +units=m +datum=WGS84 +e
llps=WGS84 +towgs84=0,0,0

Data:
                                 FILES QUANTITY CATEGORY
B1_dn LT51700721984174XXX08_B1.TIF        dn    image
B2_dn LT51700721984174XXX08_B2.TIF        dn    image
B3_dn LT51700721984174XXX08_B3.TIF        dn    image
B4_dn LT51700721984174XXX08_B4.TIF        dn    image
B5_dn LT51700721984174XXX08_B5.TIF        dn    image
B6_dn LT51700721984174XXX08_B6.TIF        dn    image
B7_dn LT51700721984174XXX08_B7.TIF        dn    image

Available calibration parameters (gain and offset):
        dn -> radiance (toa)
        dn -> brightness temperature (toa)
```

Next, we need to read the Landsat MTL metadata files, and then combine them into a single Landsat image stack based on the ***stackMeta()*** function. The ***stackMeta()*** function from the "RStoolbox" package is used to import separate Landsat files into a single stack. The ***stackMeta()*** reads Landsat MTL or XML metadata files and loads single Landsat Tiffs into a rasterStack. In order to do that, we need to create an object "**TM522_6_84**" which will store the single Landsat imagery as a raster stack as shown below.

```
TM522_6_84 <- stackMeta(metaData)
```

Check the attributes of the object "TM522_6_84", which contains the Landsat 5 TM imagery. Type "TM522_6_84" and run.

```
> TM522_6_84
```

The output is shown below.

```
class: RasterStack
dimensions: 1972, 2203, 4344316, 7  (nrow, ncol, ncell, nlayers)
resolution  : 30, 30  (x, y)
extent: 255315, 321405, -1999725, -1940565  (xmin, xmax, ymin, ymax)
coord. ref. : +proj=utm +zone=36 +datum=WGS84 +units=m +no_defs +ell
ps=WGS84 +towgs84
names       : B1_dn, B2_dn, B3_dn, B4_dn, B5_dn, B6_dn, B7_dn
min values  :    40,    14,    10,     6,     1,   112,     1
max values  :   230,   119,   138,   124,   255,   160,   255
```

The output shows "class", which is a RasterStack that was created from the raster file metadata. By definition a RasterStack is a collection of RasterLayer objects with the same spatial extent and resolution. The "dimensions" retrieve spatial dimensions from spatial data, while "resolution" provides the spatial resolution of the Landsat 5 TM imagery. The "extent" and the "coord.ref" retain the spatial extent and coordinates of the Landsat 5 TM imagery, respectively. The "names" retains the Landsat 5 TM bands, while the "min values" and "max values" provides minimum and maximum DN values of each band. Note that the "extent" (255315, 321405, −1999725, −1940565) and "coord.ref" (WGS UTM Zone 36) shown in the metadata suggests that the Landsat 5 TM imagery was acquired in the northern hemisphere. This needs to be corrected later.

Run the ***dataType()*** function in order to check the data type.

```
> dataType(TM522_6_84)[[1]]
```

```
[1] "INT2U"
```

The output shows that data type is an integer. Keep this in mind because we will need to compare the original Landsat 5 TM imagery with the radiometric-corrected Landsat 5 TM imagery later.

In Sect. 1.2.5, we talked briefly about data types. Here, we are going to discuss briefly about numerical data types such as integers and float (double), which we use in this workbook. Integers represent whole numbers, while float (double) represent continuous numbers or numbers with decimals. Data types are important since they influence the file size. Table 2.2 shows the data types that are defined in the **raster** package.

Table 2.2 Important data types

Data type	Minimum value	Minimum value
INT1S	−127	127
INT1U	0	255
INT2S	*−32,767*	*32,767*
INT2U	*0*	*65,534*
FLT4S	*−3.4e+38*	*3.4e+38*
FLT8S	*−1.7e+308*	*1.7e+308*

Step 6: Display Landsat imagery

Next, remove band 6 (the thermal band) using the ***dropLayer()*** function, which removes a layer from a RasterStack or RasterBrick. We are not going to use band 6 for atmospheric correction and classification. The ***dropLayer()*** function is available from "raster" package.

```
> TM5_22_6_84 <- dropLayer(TM522_6_84, c(6))
```

Check the attributes of the original Landsat 5 TM raster stack. Type in the script editor "TM5_22_6_84" and run.

```
> TM5_22_6_84
```

```
class       : RasterStack
dimensions: 1972, 2203, 4344316, 6  (nrow, ncol, ncell, nlayers)
resolution  : 30, 30  (x, y)
extent: 255315, 321405, -1999725, -1940565  (xmin, xmax, ymin, ymax)
coord. ref. : +proj=utm +zone=36 +datum=WGS84 +units=m +no_defs +ell
ps=WGS84 +towgs84
names       : B1_dn, B2_dn, B3_dn, B4_dn, B5_dn, B7_dn
min values  :    40,    14,    10,     6,     1,     1
max values  :   230,   119,   138,   124,   255,   255
```

The output shows that band 6 (thermal band) has been removed. Next, display the six remaining Landsat 5 TM imagery bands using the ***spplot()*** function. I have used the the ***spplot()*** function instead of the the ***plot()*** function because it easier to map several layers with a single legend. The ***spplot()*** function is a wrapper function from the **sp** package.

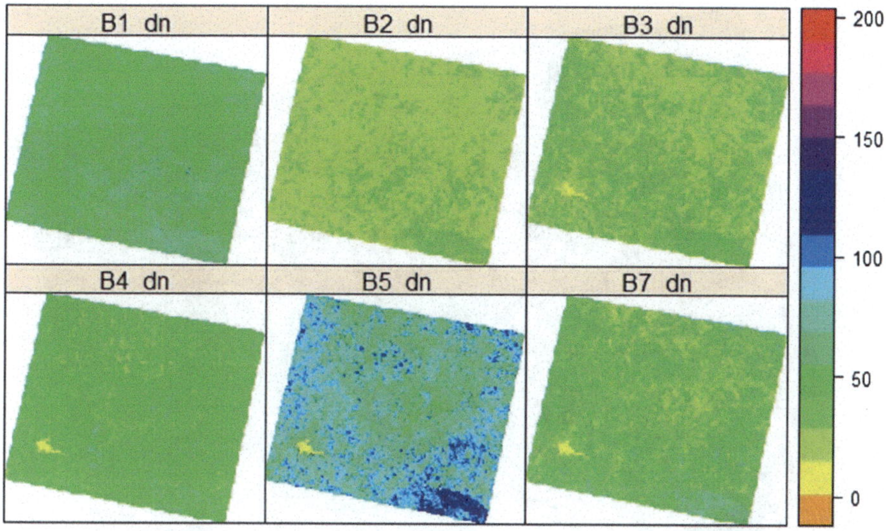

Fig. 2.5 Original Landsat 5 TM imagery acquired on 22 June 1984

```
> spplot(TM5_22_6_84, col.regions = rainbow(99, start=.
1))
```

Figure 2.5 shows the output, which comprises subsets of Landsat 5 TM band 1 (B1_dn), band 2 (B2_dn), band 3 (B3_dn), band 4 (B4_dn), band 5 (B5_dn), band 7 (B7_dn).

Note that in R, there are different plotting functions such as ***plot()***, ***ssplot()*** and ***ggr()*** that we can use to display raster data. You can use any of these plotting functions depending on how you want to display your data. It also important to note that these functions have arguments. Therefore, you can check the arguments of the functions in the help pages in R. In order to see the arguments of a function, simply type "??spplot" in the console.

Finally, display the multispectral Landsat 5 TM imagery using the ***plotRGB()*** function. The ***plotRGB()*** function can be used to create "true" or "false" color images from Landsat imagery and other satellite imagery. This produces a Red-Green-Blue (RGB) plot based

Fig. 2.6 Original Landsat 5 TM imagery acquired on 22 June 1984

on three layers (in a RasterBrick or RasterStack).

```
> plotRGB(TM5_22_6_84, r=5, g=4, b=3, stretch="lin")
```

Figure 2.6. shows the Landsat 5 TM imagery displayed as Red (band 5), Green (band 4) and Blue (band 3).

Step 7: Visualization

Graphs provide a useful way to understand data. Here, we are going to use ***ggplot2()*** function in order to create a combined histogram and density plot of the original Landsat 5 TM imagery.

Plot the histogram and density plots of the six original Landsat 5 TM imagery bands using ***ggplot()*** function. The ***ggplot2()*** function is used to create elegant data visualizations using the Grammar of Graphics. Note that I have included the **proc.time**[1] command to measure the processing time.

```
> timeStart<- proc.time() # allows to measure computation time
> B1_dn <- ggplot(TM5_22_6_84, aes(B1_dn)) +geom_histogram(aes(y = ..
density..)) + geom_density()
> B2_dn <- ggplot(TM5_22_6_84, aes(B2_dn)) +geom_histogram(aes(y = ..
density..)) + geom_density()
> B3_dn <- ggplot(TM5_22_6_84, aes(B3_dn)) +geom_histogram(aes(y = ..
density..)) + geom_density()
> B4_dn <- ggplot(TM5_22_6_84, aes(B4_dn)) +geom_histogram(aes(y = ..
density..)) + geom_density()
> B5_dn <- ggplot(TM5_22_6_84, aes(B5_dn)) +geom_histogram(aes(y = ..
density..)) + geom_density()
> B7_dn <- ggplot(TM5_22_6_84, aes(B7_dn)) +geom_histogram(aes(y = ..
density..)) + geom_density()
> proc.time() - timeStart # user time and system time.
```

```
user   system  elapsed
16.08  1.63    17.5
```

Run the ***grid.arrange()*** function in order to arrange the plots for the original Landsat 5 TM imagery bands 1, 2, 3, 4, 5 and 7 (Fig. 2.7).

---

[1]

*The **proc.time** command*

*While code execution depends on your operating system and computer specifications, it is considered a good practice to time your code that you will share. In this workbook, we use the **proc.time** as a timer. The command determines how much real and CPU time (in seconds) the running process has taken. In other words, you start with a starting time, run all the code, and then stop it by subtracting the starting time from the ending time. The **proc.time** commands prints "user", "system", and "elapsed" values. The "user" time defines the execution of the code, the "system" time relates to system processes (e.g., opening and closing files), and the "elapsed" time is the difference in times since you started the timer.*

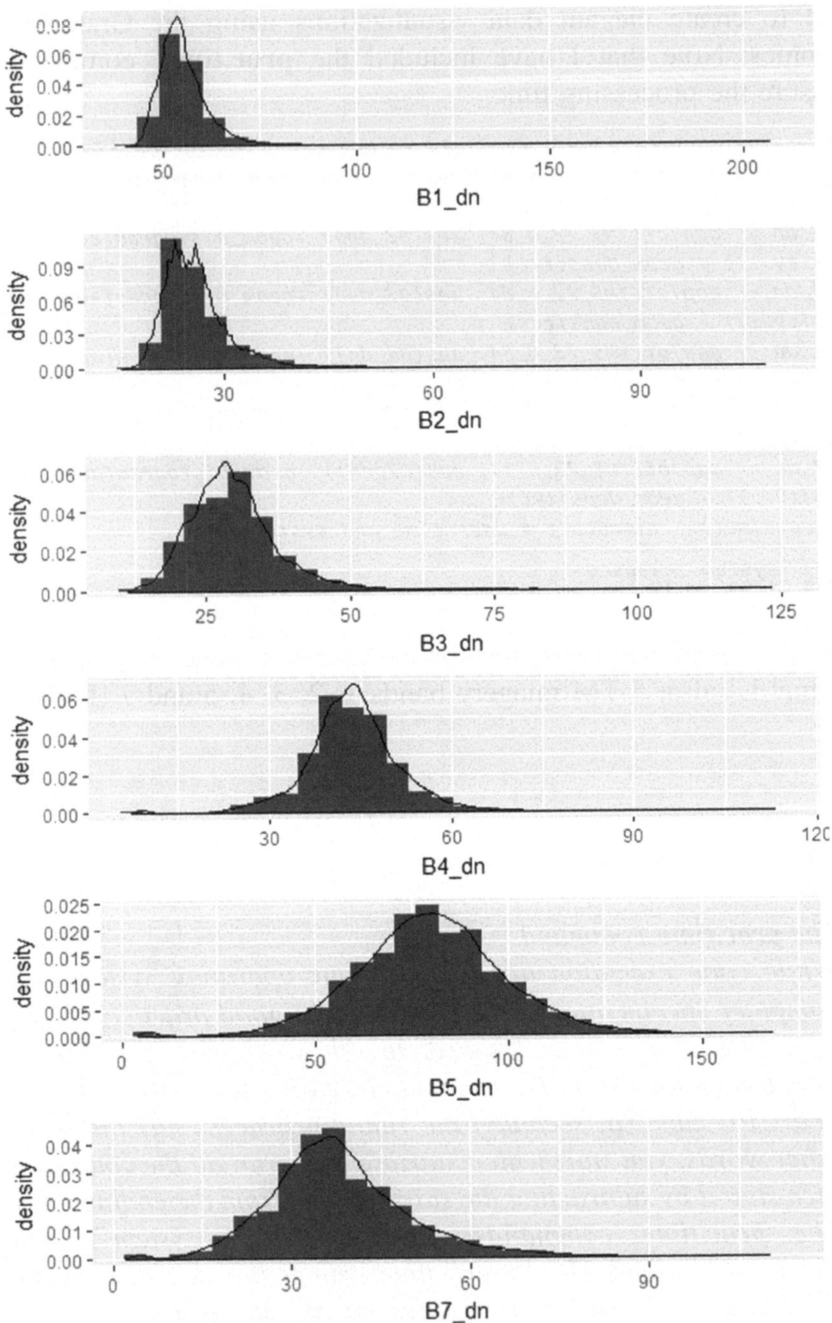

Fig. 2.7 Histogram and density plot for the original Landsat 5 TM imagery acquired on 22 June 1984

```
> grid.arrange(B1_dn, B2_dn, B3_dn, B4_dn, B5_dn, B7_dn,
  ncol = 1)
`stat_bin()` using `bins = 30`. Pick better value with
`binwidth`.There were 12 warnings (use warnings() to
see them
```

The warning message states that ***ggplot2()*** *function removed rows containing non-finite values for the histogram (stat_bin) and density plot (stat_density).*

The histogram and density plot are important graphic representations, which are useful for exploring the distribution of digital number values (e.g., normal distribution, negative or positively skewed distribution, multimodal or uniform distribution), contrast (high or low) as well as identifying bad data values in the remotely sensed imagery. Therefore, it is important to check the histogram and density plot of the remotely-sensed data given that most statistical tests used for image analysis assume that the digital values recorded in the scene are normally distributed, which is not always true (Jensen 2005). In this workbook, the ***ggplot2()*** function is used to create the histogram and density plot instead of the ***hist()*** and ***density()*** functions. This is because the ***hist()*** function takes more time to plot the histogram. Figure 2.3 shows histogram and density plots for the original Landsat 5 TM imagery. The histogram for bands 1, 2 and 3 are concentrated into the lower half of the 0–255 range (that is a positively skewed distribution), which suggests that these bands are low in contrast. However, band 5 is slightly centered.

Hint*: NoData and bad data values in raster images*
A NoDataValue in a raster image is a value assigned to pixels where data is missing or was not collected. It is assigned as NA (Not Available). In contrast, bad data values are values that fall outside of the applicable range of a dataset, which could be a result of error during either data collection or processing. Therefore, it is important to check your raster data sets before performing image processing.

2.3 Tutorial 2: Radiometric Correction and Reprojection

2.3.1 Objectives

- Perform radiometric correction for Landsat 5 TM imagery acquired on 22 June 1984,
- Explore statistics for the radiometric-corrected Landsat 5 TM imagery, and
- Reproject the COST-corrected Landsat 5 TM imagery.

2.3.2 Procedure

Note that you do not need to load the **sp**, **rgdal**, **raster**, and **RStoolbox** packages if you are still using the same script and environment as tutorial 1. However, if you closed the previous R session, you need to load the packages using the ***library()*** function as shown in step 1.

In this tutorial exercise, we are going to perform radiometric correction using models available in the **RStoolbox** package (Leutner and Horning 2016). In this workbook, radiometric correction refers to conversion from satellite digital number (DN) to radiance, and atmospheric correction. First, we will convert the Landsat DN to at-satellite radiance. Second, we will convert at-satellite radiances to top-of-the-atmosphere reflectance (that is apparent reflectance) using the apparent reflectance model. Third, we will perform radiometric correction using Chaves' simple dark object subtraction (SDOS), which requires haze correction. This will be followed by Chaves' dark object subtraction (DOS), which assumes that scattering is highest in the blue band and decreases towards the NIR. Last but not least, we will perform radiometric correction using Chaves's COST model.

As we observed in tutorial 1, the Landsat 5 TM imagery appear to have been acquired in the northern hemisphere. However, the test site is located in the southern hemisphere. Therefore, the COST-corrected Landsat 5 TM imagery should be reprojected from UTM Zone 36 N to UTM Zone 36 S. We are going to perform radiometric correction first, and then reprojection later.

A. ***Radiometric correction***

The **radCor** (Radiometric Calibration and Correction) function from **RSToolbox** package will be used to implement methods for radiometric calibration and correction of Landsat 5 TM imagery.

(i) ***Top-of-the atmosphere radiance***

Landsat sensors record the intensity of electromagnetic radiation (ER) from the Earth's surface as a digital number (DN) for each spectral band. The DN for each spectral band is a linearly transformed representation of **at-sensor radiance** for the recorded part of the Earth's surface. For example, Landsat 5 TM imagery that we are using in this workbook was downloaded in DNs of 8-bit radiometric resolution. The Landsat 5 TM imagery is stored as DNs in order to reduce the file size. Therefore, we need to use sensor and band specific parameters (gain and offset) in order to compute **at-sensor radiance**. Note that radiance—the radiation leaving the Earth (that is, upwelling towards the sensor)—is measured in watts per meter squared per steradian ($Wm^{-2}\ sr^{-1}$) (Jensen 2005). The gain and offset values for each spectral band are provided with Landsat 5 TM imagery. Following are steps to convert Landsat 5 TM imagery DNs to top-of-the atmosphere radiance.

Step 1: Check the conversion parameters (gain and offset)

First, let's check the conversion parameters (that is the gain and offset), which are in the Landsat 5 TM metadata.

```
> TM5_radParameters <- metaData$CALRAD

> TM5_radParameters
       offset  gain
B1_dn -2.191 0.671
B2_dn -4.162 1.322
B3_dn -2.214 1.044
B4_dn -2.386 0.876
B5_dn -0.490 0.120
B6_dn  1.182 0.055
B7_dn -0.216 0.066
```

The output shows the offset and gain parameters from the metadata.

Step 2: Convert Landsat 5 TM DN to top-of-the-atmosphere radiance

Next, we are going to convert Landsat 5 TM imagery DN to top-of-the-atmosphere radiance using the ***radCor()*** function, which implements different methods for radiometric calibration and correction. The user can either specify a metadata file or supply the values manually. In this step, we define the method as ***rad***.

```
> rad_22_6_84 <- radCor(TM5_22_6_84, metaData
= metaData, method = "rad")
```

Step 3: Check the attributes of the top-of-the-atmosphere radiance imagery

Let's check the attributes of the top-of-the-atmosphere radiance adjusted Landsat 5 TM imagery.

```
> rad_22_6_84
```

```
class       : RasterStack
dimensions  : 1972, 2203, 4344316, 6  (nrow, ncol, ncell, nlayers)
resolution  : 30, 30  (x, y)
extent  255315, 321405, -1999725, -1940565  (xmin, xmax, ymin, ymax)
coord. ref. : +proj=utm +zone=36 +datum=WGS84 +units=m +no_defs +ell
ps=WGS84 +towgs84
names       :  B1_tra,  B2_tra,  B3_tra,  B4_tra,  B5_tra,  B7_tra
min values  :  24.649,  14.346,   8.226,   2.870,   0.000,   0.000
max values  : 152.139, 153.156, 141.858, 106.238,  30.110,  16.614
```

The Landsat 5 TM imagery digital values are now in radiance (that is in watts per meter squared per steradian (Wm^{-2} sr^{-1}).

You can also use the ***dataType()*** function in order to check the data type. Note that the data type is now float.

```
> dataType(rad_22_6_84)[[1]]
```

```
[1] "FLT4S"
```

Step 4: Visualization

Next, let's plot the histogram and density plots of the six top-of-the-atmosphere radiance adjusted Landsat 5 TM bands using the ***ggplot()*** function.

```
> timeStart<- proc.time() # allows to measure computation time
> B1_rad <- ggplot(rad_22_6_84, aes(B1_tra)) +geom_histogram(aes(y =
  ..density..)) + geom_density()
> B2_rad <- ggplot(rad_22_6_84, aes(B2_tra)) +geom_histogram(aes(y =
  ..density..)) + geom_density()
> B3_rad <- ggplot(rad_22_6_84, aes(B3_tra)) +geom_histogram(aes(y =
  ..density..)) + geom_density()
> B4_rad <- ggplot(rad_22_6_84, aes(B4_tra)) +geom_histogram(aes(y =
  ..density..)) + geom_density()
> B5_rad <- ggplot(rad_22_6_84, aes(B5_tra)) +geom_histogram(aes(y =
  ..density..)) + geom_density()
> B7_rad <- ggplot(rad_22_6_84, aes(B7_tra)) +geom_histogram(aes(y =
  ..density..)) + geom_density()
> proc.time() - timeStart # user time and system time.
  user        system     elapsed
  3.39        0.24        3.62
```

Arrange and display the histogram and density plots for bands 1, 2, 3, 4, 5 and 7 (Fig. 2.8).

```
> grid.arrange(B1_rad, B2_rad, B3_rad, B4_rad, B5_rad,
B7_rad)
```

The warning message is just stating that ***ggplot2()*** removed rows containing non-finite values for the histogram (stat_bin) and density plot (stat_density).

Figure 2.8 shows that bands 1, 2 and 3 are partly skewed, while band 5 is slightly centered. The same pattern was observed in the histogram and density plot of the original Landsat 5 TM imagery (Fig. 2.7).

(ii) ***Top-of-the-atmosphere reflectance (apparent reflectance)***

Next, let's convert at-sensor radiance to apparent at-sensor spectral reflectance, which takes into account temporal changes in solar illumination due to Earth-Sun geometry. Reflectance is a unitless or dimensionless ratio of the radiance emittance of an object and irradiance (Mather and Koch 2011). Radiance emittance refers to the energy that flows away from the surface (e.g., thermal energy emitted by the Earth), while irradiance is the radiant energy that falls on a

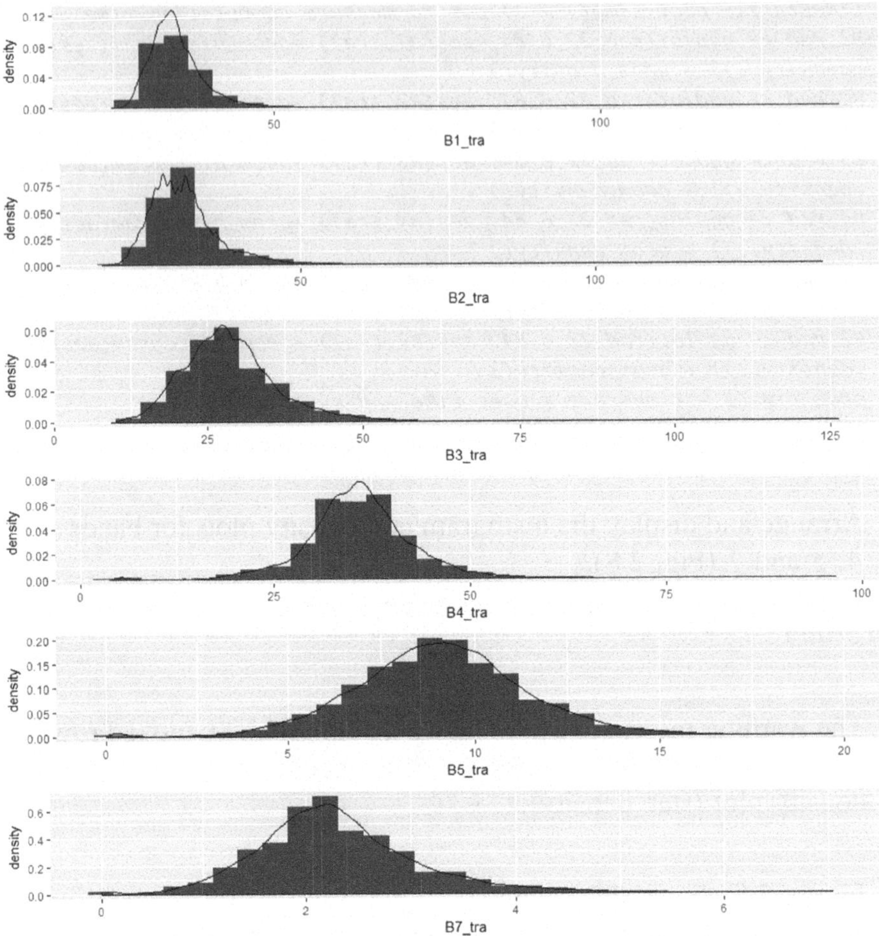

Fig. 2.8 Histogram and density plots for the top-of-the-atmosphere radiance adjusted Landsat 5 TM imagery acquired on 22 June 1984. *Note: The unit of the top-of-atmosphere radiance is* Wm^{-2} sr^{-1}

surface (Mather and Koch 2011). Following are steps to convert Landsat 5 TM imagery DNs to top-of-the atmosphere reflectance.

Step 1: Convert Landsat 5 TM image digital numbers to top-of-the-atmosphere reflectance

We are going to use the ***radCor()*** function and ***apref*** method to compute apparent reflectance (top-of-atmosphere reflectance) as follows.

```
> apref_22_6_84 <- radCor(TM5_22_6_84, metaData =
metaData, method = "apref")
```

Step 2: Check the attributes of the apparent reflectance-corrected Landsat 5 TM image

Next, check the attributes of the apparent reflectance-corrected Landsat 5 TM imagery.

```
> apref_22_6_84
class       : RasterStack
dimensions  : 1972, 2203, 4344316, 6  (nrow, ncol, ncell, nlayers)
resolution  : 30, 30  (x, y)
extent: 255315, 321405, -1999725, -1940565  (xmin, xmax, ymin, ymax)
coord. ref. : +proj=utm +zone=36 +datum=WGS84 +units=m +no_defs +ell
ps=WGS84 +towgs84
names:        B1_tre,  B2_tre,  B3_tre,  B4_tre,
min values:  0.0715,  0.04455, 0.0301,  0.0157,
max values:  0.4405,  0.47525, 0.5185,  0.5814
```

The output shows that the minimum and maximum values are now between 0 and 1, which are the reflectance values.

Step 3: Visualization

Let's display the apparent reflectance-corrected Landsat 5 TM imagery (Fig. 2.9).

```
> plotRGB(apref_22_6_84, r=5, g=4, b=3, stretch="lin")
```

Next, let's plot the histogram and density plots of the apparent reflectance-corrected Landsat 5 TM bands using ***ggplot()*** function.

```
> B1_apref <- ggplot(apref_22_6_84, aes(B1_tre)) +geom_histogram(aes
(y = ..density..)) + geom_density()
> B2_apref <- ggplot(apref_22_6_84, aes(B2_tre)) +geom_histogram(aes
(y = ..density..)) + geom_density()
> B3_apref <- ggplot(apref_22_6_84, aes(B3_tre)) +geom_histogram(aes
(y = ..density..)) + geom_density()
> B4_apref <- ggplot(apref_22_6_84, aes(B4_tre)) +geom_histogram(aes
(y = ..density..)) + geom_density()
```

Arrange the histogram and density plots for bands 1, 2, 3, and 4. Display the combined histogram and density plots (Fig. 2.10).

Fig. 2.9 Apparent reflectance-corrected Landsat 5 TM imagery

```
> grid.arrange(B1_apref, B2_apref, B3_apref,B4_apref)
```

Figure 2.10 shows the same pattern that we observed in Figs. 2.7 and 2.4. However, the reflectance values are now between 0 and 1.

(iii) ***Simple dark object subtraction (SDOS)***

Generally, it is assumed that any radiation originating from dark pixels (that is, reflectance close to zero) in the image is due to atmospheric haze and not the reflectance of the surface. Therefore, haze values are estimates of path radiance. Let's perform radiometric correction using the simple dark object subtraction (Chavez decay model). Note that SDOS requires the estimation of haze values in the visible and NIR bands. This is because atmospheric haze affects mostly the visible wavelength range (blue, green and red bands). In

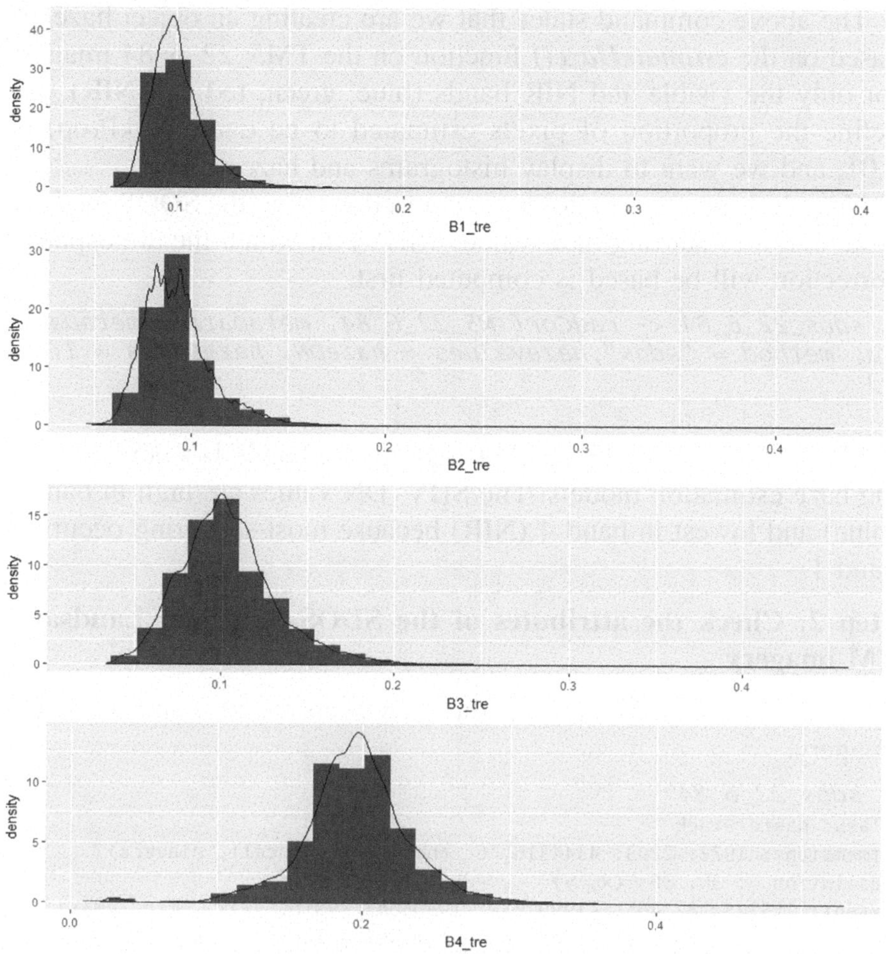

Fig. 2.10 Histogram and density plots for the apparent reflectance-corrected Landsat 5 TM imagery

this workbook, we are also going to include the NIR band for haze correction. Following are key steps for the SDOS model.

Step 1: Haze correction

First, perform automatic haze estimation using the ***estimateHaze()*** function, which estimates the DN pixel value of dark objects for the visible and NIR bands.

```
> hazeDN <- estimateHaze(TM5_22_6_84, hazeBands = 1:4,
darkProp = 0.01, plot = TRUE)
```

The above command states that we are creating an object hazeDN based on the ***estimateHaze()*** function on the TM5_22_6_84 imagery for only the visible and NIR bands (blue, green, red and NIR). We define the proportion of pixels estimated to be dark (darkProp) as 0.01, and we want to display histograms and haze values.

Next, let's perform radiometric correction using the SDOS method. Note that the "starting haze value" (SHV) on which the atmospheric correction will be based is computed first.

```
> sdos_22_6_84 <- radCor(TM5_22_6_84, metaData = metaDa
ta, method = "sdos",hazeValues = hazeDN, hazeBands = 1:
4)
```

Figure 2.11 shows the band histograms for bands 1–4 derived from the haze estimation models. The SHV_DN values are high in band 1 (blue) and lowest in band 4 (NIR) because most scattering occurs in band 1.

Step 2: Check the attributes of the SDOS-corrected Landsat 5 TM imagery

Let's check the attributes of the SDOS-corrected Landsat 5 TM imagery.

```
> sdos_22_6_84
class: RasterStack
dimensions: 1972, 2203, 4344316, 6  (nrow, ncol, ncell, nlayers)
resolution  : 30, 30  (x, y)
extent: 255315, 321405, -1999725, -1940565  (xmin, xmax, ymin, ymax)
coord. ref. : +proj=utm +zone=36 +datum=WGS84 +units=m +no_defs +ell
ps=WGS84 +towgs84=0,0,0
names:         B1_sre,     B2_sre,      B3_sre,      B4_sre,
min values:    0.0042,     0.0018,      0.0000,      0.0004
max values:    0.3733,      0.4325,     0.4870,      0.5660
```

Note that the band value ranges are now between 0 and 1.

Step 3: Visualization

Let's display the SDOS-corrected Landsat 5 TM imagery (Fig. 2.12).

```
> plotRGB(sdos_22_6_84, r=5, g=4, b=3, stretch="lin")
```

Next, let's plot the histogram and density plots of the SDOS-corrected Landsat 5 TM imagery using ***ggplot()***.

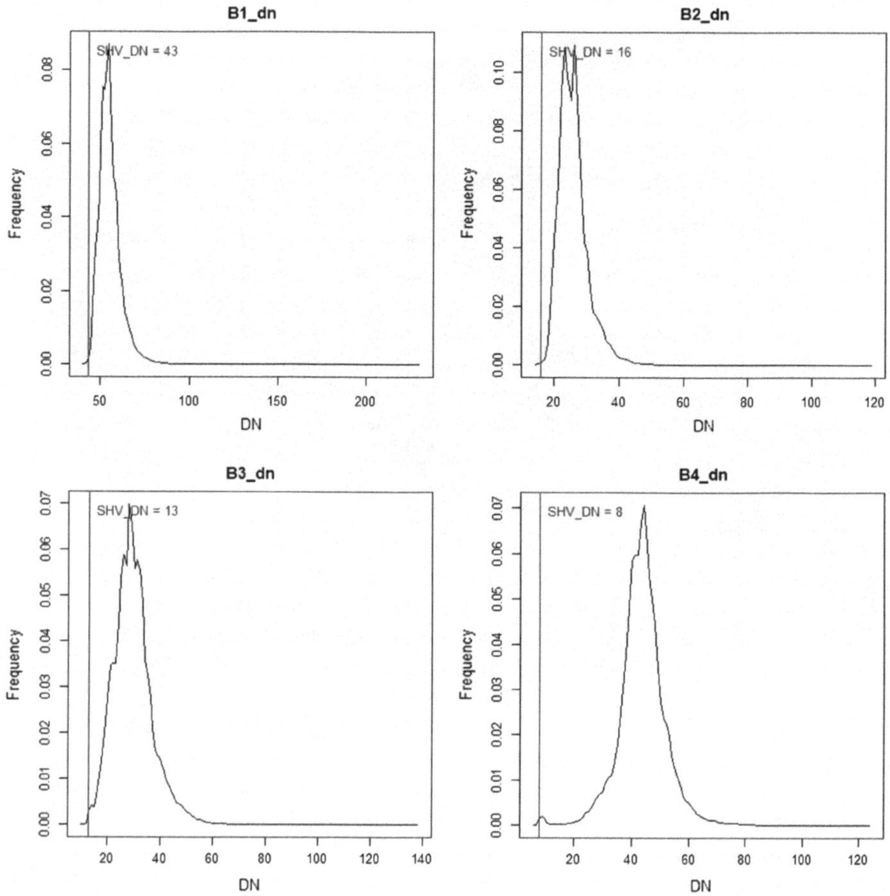

Fig. 2.11 Haze estimation models for SDOS. The horizontal axis shows the DN, while the vertical axis shows the frequency

```
> B1_sdos <- ggplot(sdos_22_6_84, aes(B1_sre)) +geom_histogram(aes(y
  = ..density..)) + geom_density()
> B2_sdos <- ggplot(sdos_22_6_84, aes(B2_sre)) +geom_histogram(aes(y
  = ..density..)) + geom_density()
> B3_sdos <- ggplot(sdos_22_6_84, aes(B3_sre)) +geom_histogram(aes(y
  = ..density..)) + geom_density()
> B4_sdos <- ggplot(sdos_22_6_84, aes(B4_sre)) +geom_histogram(aes(y
  = ..density..)) + geom_density()
```

Arrange and display the histogram and density plots for bands 1, 2, 3, and 4 (Fig. 2.13).

Fig. 2.12 SDOS-corrected Landsat 5 TM imagery acquired on 22 June 1984

```
> grid.arrange(B1_sdos, B2_sdos, B3_sdos, B4_sdos, ncol
= 1)
```

While there are changes in the reflectance values, Fig. 2.13 shows the same distribution pattern as the apparent reflectance-corrected Landsat 5 TM histograms (Fig. 2.10).

(iv) ***Dark object subtraction (DOS)***

The DOS procedure uses the atmospheric haze decay model, which assumes that scattering is highest in the blue wavelength and decreases towards the NIR. Following are steps for the DOS model.

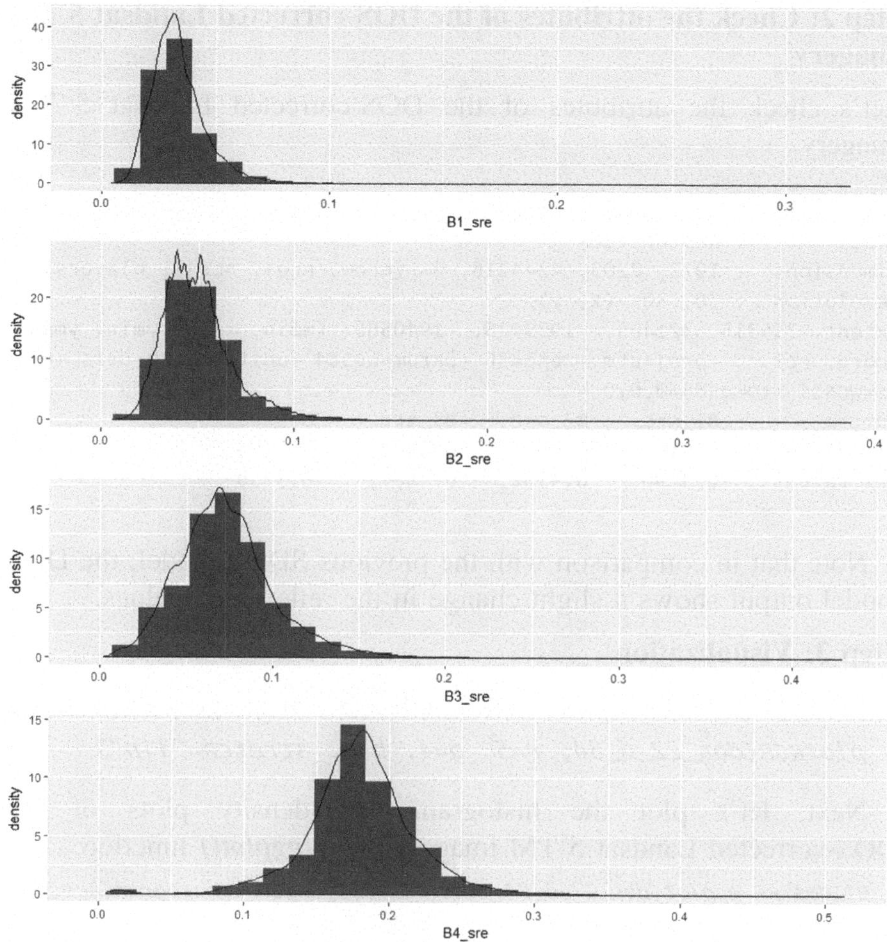

Fig. 2.13 Histogram and density plots for the SDOS-corrected Landsat 5 TM imagery acquired on 22 June 1984

Step 1: Perform "dark object subtraction" (DOS)

Execute the command below to perform atmospheric correction using the DOS model (Chavez decay model).

```
> dos_22_6_84 <- radCor(TM5_22_6_84, metaData = metaDat
a, method = "dos")
```

Step 2: Check the attributes of the DOS-corrected Landsat 5 TM imagery

Let's check the attributes of the DOS-corrected Landsat 5 TM imagery.

```
> dos_22_6_84
class        : RasterStack
dimensions   : 1972, 2203, 4344316, 6  (nrow, ncol, ncell, nlayers)
resolution   : 30, 30  (x, y)
extent: 255315, 321405, -1999725, -1940565  (xmin, xmax, ymin, ymax)
coord. ref. : +proj=utm +zone=36 +datum=WGS84 +units=m +no_defs +ell
ps=WGS84 +towgs84=0,0,0
names:          B1_sre,   B2_sre,   B3_sre,    B4_sre,
min values:     0.0105,   0.0000,   0.0000,    0.0103
max values:     0.3796,   0.4132,   0.4881,    0.5759
```

Note that in comparison with the previous SDOS model, the DOS model output shows a slight change in the reflectance values.

Step 3: Visualization

Let's display the DOS-corrected Landsat 5 TM imagery (Fig. 2.14).

```
> plotRGB(dos_22_6_84, r=5, g=4, b=3, stretch="lin")
```

Next, let's plot the histogram and density plots of the DOS-corrected Landsat 5 TM imagery using ***ggplot()*** function.

```
> B1_dos <- ggplot(dos_22_6_84, aes(B1_sre)) +geom_histogram(aes(y =
  ..density..)) + geom_density()
> B2_dos <- ggplot(dos_22_6_84, aes(B2_sre)) +geom_histogram(aes(y =
  ..density..)) + geom_density()
> B3_dos <- ggplot(dos_22_6_84, aes(B3_sre)) +geom_histogram(aes(y =
  ..density..)) + geom_density()
> B4_dos <- ggplot(dos_22_6_84, aes(B4_sre)) +geom_histogram(aes(y =
  ..density..)) + geom_density()
```

Arrange and display the histogram and density plots for bands 1, 2, 3, and 4 (Fig. 2.15).

```
> grid.arrange(B1_dos, B2_dos, B3_dos, B4_dos,ncol = 1)
```

Fig. 2.14 DOS-corrected Landsat 5 TM imagery acquired on 22 June 1984

(v) *Cosine estimation of atmospheric transmittance (COST)*

Finally, let's perform atmospheric correction using the COST model (Chavez decay model). Following are steps for the COST model.

Step 1: Perform COST radiometric correction

First, let's perform automatic haze estimation using the ***estimateHaze ()*** function (Fig. 2.16).

```
> hazeDN <- estimateHaze(TM5_22_6_84, hazeBands = 1:4,
darkProp = 0.01, plot = TRUE)
```

Next, let's perform radiometric correction by specifying the method as costz as shown in the command below.

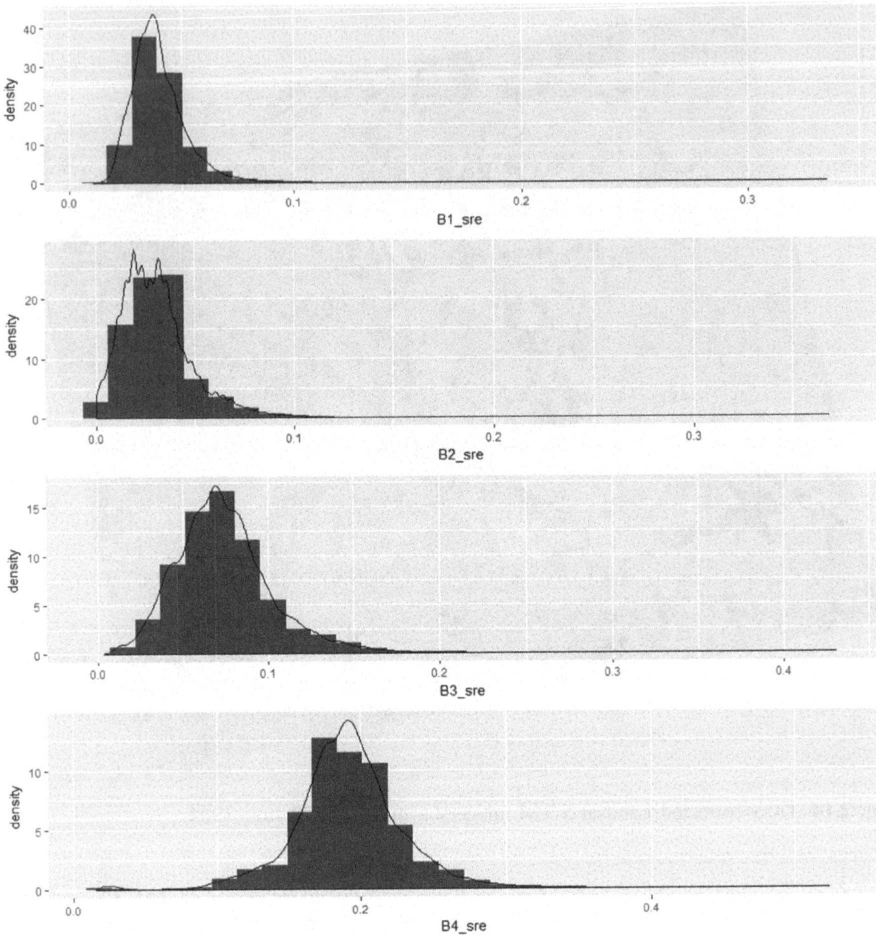

Fig. 2.15 Histogram and density plots for the DOS-corrected Landsat 5 TM imagery acquired on 22 June 1984

```
> costz_22_6_84 <-radCor(TM5_22_6_84, metaData=metaData,
  method="costz",hazeValues=hazeDN, hazeBands = 1:4)
```

Ignore the Warning message that appears!!!

Step 2: Check the attributes of the COST-corrected Landsat 5 TM imagery

Let's check the attributes of the COST-corrected Landsat 5 TM imagery.

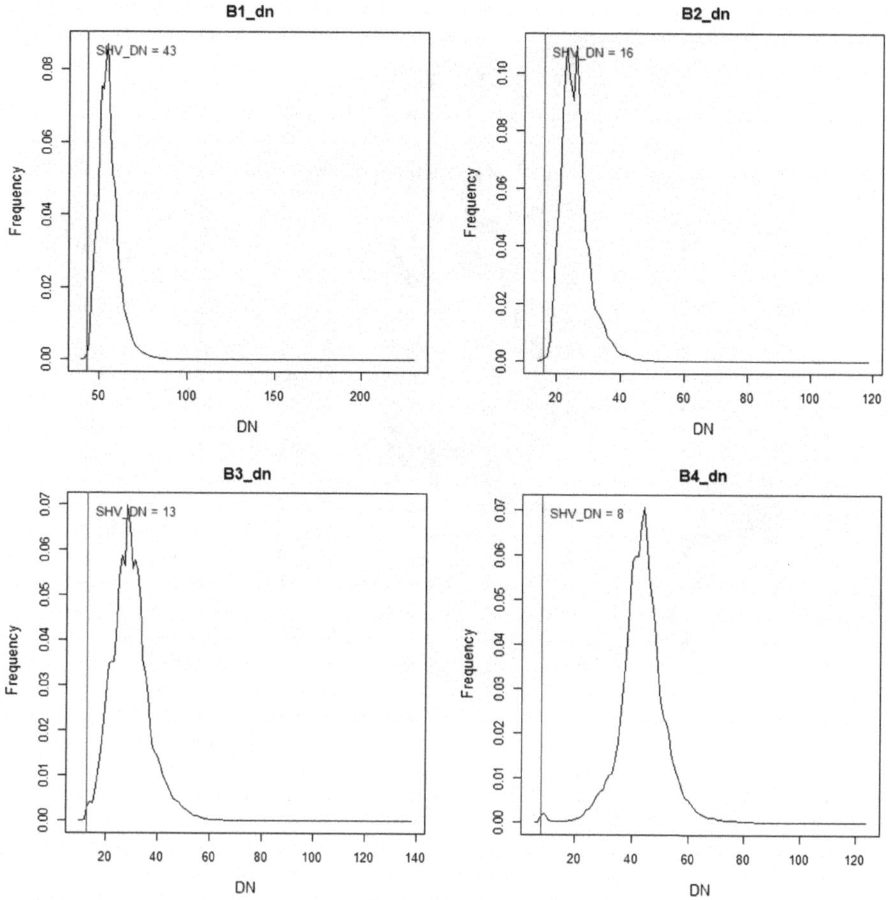

Fig. 2.16 Haze estimation models for COST

```
> costz_22_6_84
class       : RasterStack
dimensions: 1972, 2203, 4344316, 6  (nrow, ncol, ncell, nlayers)
resolution  : 30, 30  (x, y)
extent: 255315, 321405, -1999725, -1940565  (xmin, xmax, ymin, ymax)
coord. ref. : +proj=utm +zone=36 +datum=WGS84 +units=m +no_defs +ell
ps=WGS84 +towgs84=0,0,0
names:          B1_sre,    B2_sre,     B3_sre,     B4_sre
min values  : 0.01090,    0.0000,     0.0000,     0.0160
max values  : 0.6558,     0.7137,     0.8485,     1.0000
```

Fig. 2.17 COST-corrected Landsat 5 TM imagery acquired on 22 June 1984

The output shows that there a significant change in the maximum reflectance values of bands 1–3.

Step 3: Visualization

Let's display the COST-corrected Landsat 5 TM imagery (Fig. 2.17).

```
> plotRGB(costz_22_6_84, r=5, g=4, b=3, stretch="lin")
```

Next, let's plot the histogram and density plots of the COST-corrected Landsat 5 TM imagery.

```
> B1_costz <- ggplot(costz_22_6_84, aes(B1_sre)) +geom_histogram(aes
(y = ..density..)) + geom_density()
> B2_costz <- ggplot(costz_22_6_84, aes(B2_sre)) +geom_histogram(aes
(y = ..density..)) + geom_density()
> B3_costz <- ggplot(costz_22_6_84, aes(B3_sre)) +geom_histogram(aes
(y = ..density..)) + geom_density()
> B4_costz <- ggplot(costz_22_6_84, aes(B4_sre)) +geom_histogram(aes
(y = ..density..)) + geom_density()
```

Display the combined histogram and density plots for bands 1, 2, 3, and 4 (Fig. 2.18).

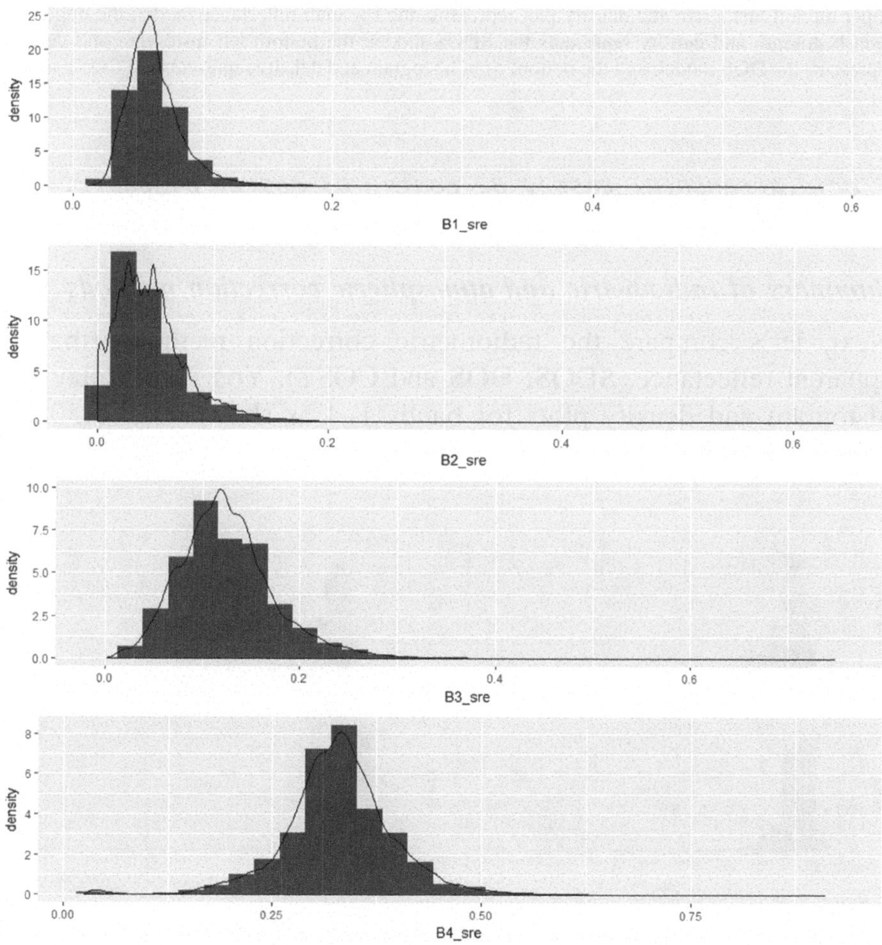

Fig. 2.18 Histogram and density plots for the COST-corrected Landsat 5 TM imagery acquired on 22 June 1984

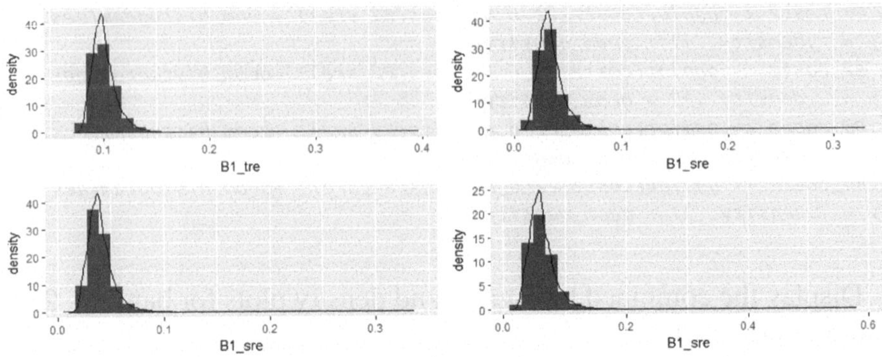

Fig. 2.19 Comparison of radiometric correction models for Landsat 5 TM band 1. *Note* The upper top left histogram and density plot represents the apparent reflectance model; the upper top right histogram and density represents the SDOS model; the bottom left histogram and density represents the DOS model; and the bottom right histogram and density represents the COST model

```
> grid.arrange(B1_costz, B2_costz, B3_costz, B4_costz,
ncol = 1)
```

Summary of radiometric and atmospheric correction methods

Next, let's compare the radiometric correction methods (that is apparent reflectance, SDOS, DOS and COST). You can display the histogram and density plots for bands 1–4 as shown in Figs. 2.19, 2.20, 2.21 and 2.22.

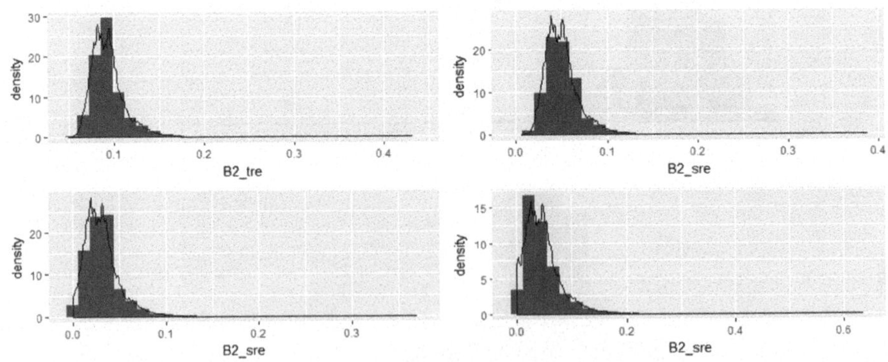

Fig. 2.20 Comparison of atmospheric correction models for Landsat 5 TM band 2

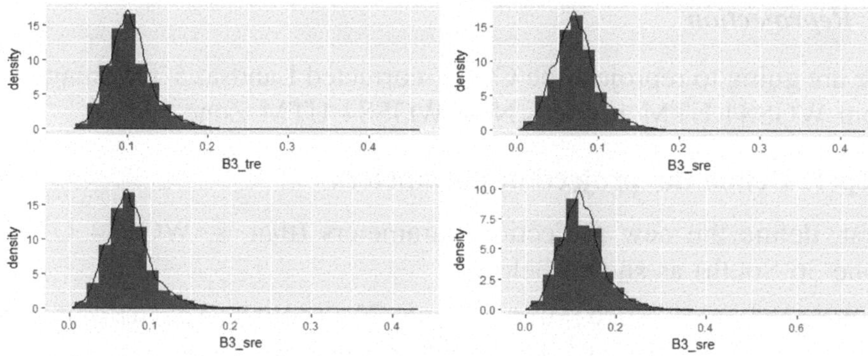

Fig. 2.21 Comparison of atmospheric correction model for Landsat 5 TM band 3

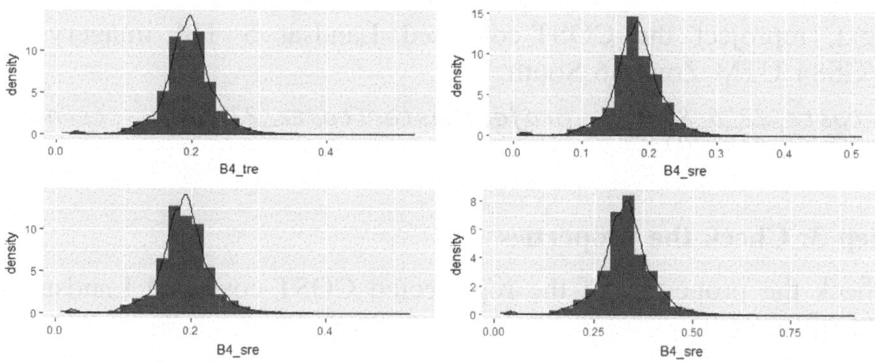

Fig. 2.22 Comparison of atmospheric correction model for Landsat 5 TM band 4

```
> grid.arrange(B1_apref, B1_sdos, B1_dos, B1_costz)
> grid.arrange(B2_apref, B2_sdos, B2_dos, B2_costz)
> grid.arrange(B3_apref, B3_sdos, B3_dos, B3_costz)
> grid.arrange(B4_apref, B4_sdos, B4_dos, B4_costz)
```

Figures 2.19, 2.20, 2.21 and 2.22 shows the radiometric-corrected Landsat 5 TM bands, which were computed using the apparent reflectance, SDOS, DOS and COST models. Note that most of the scattering is corrected in blue and green bands, and slightly in the red band. The COST-corrected Landsat image will be used for image processing and classification since the COST model accounts for multiplicative atmospheric scattering and absortiption effects.

B. *Reprojection*

We are going to reproject the COST-corrected Landsat 5 TM imagery from WGS84 UTM zone 36 N to WGS84 UTM Zone 36 S.

Step 1: Define the projection parameters

First, define the new projection parameters (that is, WGS84 UTM Zone 36 South) as shown below.

```
> newproj <- "+proj=utm +zone=36 +south +datum=WGS84
+units=m +no_defs +ellps=WGS84 +towgs84=0,0,0"
```

Step 2: Preform reprojection

Next, reproject the COST-corrected Landsat 5 TM imagery to WGS84 UTM Zone 36 South.

```
> costz_22_6_84pr <- projectRaster(costz_22_6_84, crs=n
ewproj, res=30)
```

Step 3: Check the properties

Check the properties of the reprojected COST-corrected Landsat 5 TM imagery.

```
> costz_22_6_84pr
class       : RasterBrick
dimensions  : 1982, 2213, 4386166, 6  (nrow, ncol, ncell, nlayers)
resolution  : 30, 30  (x, y)
extent  : 255165, 321555, 8000125, 8059585  (xmin, xmax, ymin, ymax)
coord. ref. : +proj=utm +zone=36 +south +datum=WGS84 +units=m +no_defs +ellps=WGS84 +towgs
4=0,0,0
data source : in memory
names: B1_sre,  B2_sre, B3_sre,  B4_sre,  B5_sre,   B7_sre
min values: 1.090038e-02, 0.00000e+00, 7.046826e-14, 1.600820e-02, 0.00000e+00, 0.00000e+0
max values: 0.6557847,    0.7136533,    0.8484794,    1.0000000,    1.0000000,    1.000000
```

Step 4: Display the reprojected COST-corrected Landsat 5 TM imagery

Display the reprojected COST-corrected Landsat 5 TM composite imagery.

```
> plotRGB(TM5_22_6_84, r=5, g=4, b=3, stretch="lin")
```

Step 5: Create a subset of the reprojected COST-corrected Landsat 5 TM imagery

Next, create a subset of the reprojected COST-corrected Landsat 5 TM imagery that corresponds to the test site. First, import the boundary shapefile.

```
> clip_boundary <- readOGR(getwd(), "Hre_Boundary")
OGR data source with driver: ESRI Shapefile
Source: "E:/~/Original_Subsets", layer: "Hre_Boundary"
with 1 features
It has 3 fields
Integer64 fields read as strings:  id area_int
```

Create a clip function in order to clip the reprojected COST-corrected Landsat 5 TM imagery.

```
> clip<-function(raster,shape) {
   a1_crop<-crop(raster,shape)
   step1<-rasterize(shape,a1_crop)
   a1_crop*step1}
```

Next, clip the reprojected COST-corrected Landsat 5 TM imagery using the clip function and the boundary shapefile for the test site.

```
> costz22_6_84 = clip(costz_22_6_84pr, clip_boundary)
```

Step 6: Check attributes

Check the attributes and dimensions of the reprojected COST-corrected Landsat 5 TM imagery subset.

```
> costz22_6_84
class       : RasterBrick
dimensions  : 1533, 1832, 2808456, 6  (nrow, ncol, ncell, nlayers)
resolution  : 30, 30  (x, y)
extent      : 262065, 317025, 8002765, 8048755  (xmin, xmax, ymin, y
max)
coord. ref. : +proj=utm +zone=36 +south +datum=WGS84 +units=m +no_de
fs +ellps=WGS84 +towgs84=0,0,0
data source : in memory
names       :     layer.1,     layer.2,     layer.3,     layer.4,
  layer.5,     layer.6
min values  : 0.010900380, 0.000000000, 0.001815953, 0.016008202, 0.
000000000, 0.000000000
max values  :   0.6557847,   0.7136533,   0.8484794,   1.0000000,
1.0000000,   1.0000000
```

Fig. 2.23 Multispectral Landsat 5 TM imagery for the test site

Step 7: Display the subset

Next, display the subset for the test site (Fig. 2.23).

```
> plotRGB(costz22_6_84, r=5, g=4, b=3, stretch="lin")
```

Step 8: Save the COST-corrected Landsat 5 TM imagery subset

Save the reprojected COST-corrected Landsat 5 TM imagery, which we are going to use for image processing and classification in Chaps. 3, 4 and 5.

```
> writeRaster(cost22_6_84,"E:\\path~toData\\CostzPr_22_
6_84.img", datatype='FLT4S', overwrite = TRUE)
```

2.4 Summary

In this chapter, we learnt how to import and check the properties of Landsat 5 TM imagery as well as creating histogram and density graphs. In addition, we learnt how to perform pre-processing procedures such as radiometric correction and reprojection. In the next chapter, you will learn about image transformation using spectral and texture indices.

2.5 Additional Exercises

(i) Import the Landsat 5 TM imagery acquired on 31 December 1984.
(ii) Compute the histogram and density plot for the Landsat 5 TM imagery acquired on 31 December 1984.
(iii) Compare the histogram and density plot for the Landsat 5 TM imagery acquired on 31 December 1984 with the Landsat 5 TM imagery acquired on 22 June 1984.
(iv) Perform radiometric correction for the Landsat 5 TM imagery acquired on 31 December 1984 using the procedures outlined in Sect. 2.3.
(v) Compare your COST-corrected Landsat 5 TM imagery acquired on 31 December 1984 with the COST-corrected Landsat 5 TM imagery acquired on 22 June 1984.
(vi) Reproject the COST-corrected Landsat 5 TM imagery acquired on 31 December 1984.

References

Chavez PS (1996) Image-based atmospheric correction-revisited and improved. Photogramm Eng Remote Sens 62:1025–1036

Du Y, Teillet PM, Cihlar J (2002) Radiometric normalization of multitemporal high-resolution satellite images with quality control for land cover change detection. Remote Sens Environ 82:123–134

Jensen JR (2005) Remote sensing of the environment: an earth resource perspective
Leutner B, Horning N (2016) Package 'RStoolbox'. http://bleutner.github.io/RStoolbox. Accessed 2 April 2017
Mather PM, Koch M (2011) Computer processing of remotely-sensed images: an introduction. Wiley-Blackwell, Chichester

Chapter 3
Image Transformation

Abstract Image transformation refers to the processing of spectral bands into "new" images in order to highlight image features of interest. Remote sensing literature review indicates that spectral and spatial indices can improve land use/cover classification accuracy. In this workbook, selected vegetation and texture indices will be computed from Landsat 5 TM imagery. The selected vegetation and texture indices will be used for image classification in Chap. 5.

Keywords Image transformation · Vegetation indices · Texture · Landsat 5 TM imagery

3.1 Background

Image transformation is the processing of spectral bands into "new" images in order to highlight image features of interest. Examples are spectral indices or band rationing, image texture, and principal components analysis. Previous remote sensing studies have reported that spectral and spatial indices improve land use/cover classification accuracy (Tso and Mather 2001; Shaban and Dikshit 2001). In this workbook, selected vegetation and texture indices will be computed from Landsat 5 TM imagery. The selected vegetation and texture indices will be used in Chap. 5. Following is a brief description of vegetation and texture indices.

Electronic supplementary material The online version of this chapter (https://doi.org/10.1007/978-981-13-8012-9_3) contains supplementary material, which is available to authorized users.

C. Kamusoko, *Remote Sensing Image Classification in R*, Springer Geography,
https://doi.org/10.1007/978-981-13-8012-9_3

3.1.1 Spectral Indices

Spectral indices are derived surface reflectance from two or more bands, which indicate relative abundance of features of interest. While there are many spectral indices in literature, in this workbook we will calculate vegetation indices such as the Normalized Difference Vegetation Index (NDVI), Soil Adjusted Vegetation Index (SAVI), and Modified Soil Adjusted Vegetation Index (MSAVI). The vegetation indices will be calculated using the **RSToolbox** package (Leutner and Horning 2016).

NDVI (Rouse et al. 1974) is a ratio between the red (R) and near infrared (NIR) values, which measures photosynthetic biomass that correlates with green vegetation cover. The NDVI values range between −1 and 1. High values indicate vegetation greenness, while low values indicate non-vegetated areas (Wegmann et al. 2016). While NDVI is one of the commonly used vegetation index, it is influenced by soil reflectance especially in areas with sparse vegetation cover. In order to correct for soil reflectance, Huete (1988) developed SAVI, which is a ratio between the Red and NIR values with a soil brightness correction factor. Later on, Qi et al. (1994) developed MSAVI to minimize soil effects. MSAVI is a ratio between the R and NIR values with an L function applied to maximize reduction of soil effects on the vegetation signal.

3.1.2 Texture Indices

Image texture refers to the spatial distribution of tonal variations or the pattern of spatial relationships among the grey levels of neighboring pixels (Mather 1999; Haralick et al. 1973). A wide range of techniques are available for computing texture indices from remotely-sensed images (Berberoglu et al. 2000). In this workbook, Grey Level Co-occurrence Matrix (GLCM) indices will be computed. GLCM is an approximation of joint probabilistic density function of pixel pairs (Herold et al. 2003). In other words, GLCM tabulates how often different combinations of pixel brightness values (grey levels) occur in an image. There are first, second, third or high order GLCM measures. The first order texture considers the original image statistics like variance, and do not consider pixel neighbor relationships. The second order measures consider the relationship between groups of two neighboring pixels in the original image. Last, third and higher order texture indices consider the relationships among three or more pixels. However, third and higher order textures take a long time to compute and interpretation is difficult. It should be noted that texture measurement requires the choice of the window size, direction of offset, offset distance, and an appropriate band.

In this workbook, we are going to compute second order GLCM indices such as the mean, variance, homogeneity and entropy from Landsat bands 4, 5 and 7 based on a 3 × 3 moving window. Landsat bands 4, 5 and 7 are selected because they provide maximum variability in terms of the range of grey scale values, which can be used to separate settlement areas and non-vegetated surfaces. The **glcm** package developed by Zvoleff (2016) will be used to compute the GLCM indices.

The mean shows the overall intensity level in the neighborhood, while variance measures heterogeneity (Herold et al. 2003). For example, variance increases when the grey level values differ from their mean. Homogeneity measures image uniformity or similarity, whereas Entropy measures the randomness of the intensity distribution. Entropy is high when an image is not texturally uniform (Herold et al. 2003; Haralick et al. 1973). Although there are many GLCM indices, most of them are highly correlated. Therefore, we are going to compute only four GLCM indices. However, readers are encouraged to test other GLCM indices in their study areas.

3.2 Tutorial 1: Vegetation Indices

3.2.1 Objective

- Compute vegetation indices using Landsat 5 TM imagery acquired on 22 June 1984.

3.2.2 Overview of the Packages

In tutorial 1, we are going to use the **sp**, **spplot**, **rgdal**, **raster**, **RStoolbox** packages that were used in the previous chapters. In addition, we are going to use **RColorBrewer** package, which provides color schemes for maps. Therefore, you need to ensure that the **RColorBrewer** package is installed.

3.2.3 Procedure

Step 1: Load the necessary packages

First, load the following packages.

```
> library(sp)
> library(rgdal)
> library(raster)
> library(ggplot2)
> library(gridExtra)
> library(RStoolbox)
> install.packages("RColorBrewer")
> library(RColorBrewer)
```

Step 2: Load the Landsat image stack

Set up the working directory.

```
setwd("c:\\path\\to\\Data\\folder")
```

Next, create a list of raster layers or bands that will be imported.

```
> rlist<-list.files(getwd(),pattern="img$",
        full.names=TRUE)
```

Combine or stack the raster layers.

```
> costz22_6_84 <- stack(rlist)
```

Step 3: Check the attributes

Check the attributes and dimensions of the Landsat 5 TM imagery.

```
> costz22_6_84
class       : RasterStack
dimensions  : 1533, 1832, 2808456, 6  (nrow, ncol, ncell, nlayers)
resolution  : 30, 30  (x, y)
extent  : 262065, 317025, 8002765, 8048755  (xmin, xmax, ymin, ymax)
```

Step 4: Display the Landsat images

Next, let's display the composite Landsat 5 TM subset for the test site.

```
> plotRGB(costz22_6_84, r=5, g=4, b=3, stretch="lin")
```

Step 5: Compute selected vegetation indices

Compute NDVI, SAVI and MSAVI vegetation indices.

```
VI_06_84 <- spectralIndices(costz22_6_84,
            red = "CostzPr_22_6_84.3",
            nir = "CostzPr_22_6_84.4",
            indices = c("NDVI", "SAVI", "MSAVI"))
```

Check the properties of the vegetation indices.

```
> VI_06_84
class       : RasterBrick
dimensions: 1533, 1832, 2808456, 3  (nrow, ncol, ncell, nlayers)
resolution  : 30, 30  (x, y)
extent  : 262065, 317025, 8002765, 8048755  (xmin, xmax, ymin, ymax)
coord. ref. : +proj=utm +zone=36 +south +ellps=WGS84 +towgs84=0,0,0,
-0,-0,-0,0 +units=m +no_defs
data source : in memory
names       :       NDVI,        SAVI,       MSAVI
min values  : -0.5661509, -0.3024195, -0.7052005
max values  :  0.9872926,  0.9089098,  0.8822286
```

The output shows that all three vegetation indices have different value ranges. Next, let's plot the vegetation indices. First, select a color brewer of your choice and then create a color palette. After that, display the vegetation indices. Following are the commands.

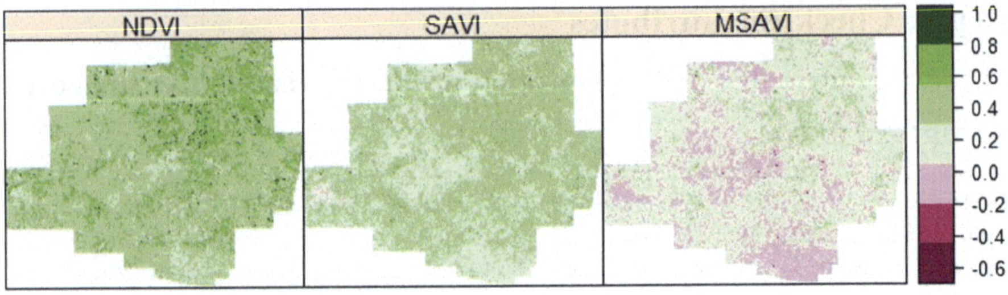

Fig. 3.1 Vegetation indices for the test site

```
display.brewer.all()

VI_palette <- brewer.pal(n = 10, name = "PiYG")

spplot(VI_06_84 , col.regions = VI_palette, cuts = 6,
col = "transparent")
```

Figure 3.1 shows differences in the vegetation indices. In this workbook, we are going to use vegetation indices computed from Landsat 5 TM imagery acquired in the dry and wet seasons. The idea is to check if the vegetation phenology can help to improve land use/ cover classification in the test site.

Finally, save the individual vegetation indices as shown in the command below. We are going to use these vegetation indices for land use/cover classification in Chap. 5.

```
>writeRaster(VI_06_84$NDVI,"G:~\\NDVI_06_84.img",datatype='FLT4S',
overwrite = TRUE)

>writeRaster(VI_06_84$SAVI,"G:~\\SAVI_06_84.img",datatype='FLT4S',
overwrite = TRUE)

>writeRaster(VI_06_84$MSAVI,"G:~MSAVI_06_84.img",datatype='FLT4S',
overwrite = TRUE)
```

3.3 Tutorial 2: Texture Analysis

3.3.1 Objective

- Compute texture indices from the Landsat 5 TM imagery acquired on 22 June 1984.

3.3.2 *Overview of the Packages*

In this tutorial, we are going to use the **sp**, **rgdal**, **raster**, **RStoolbox** packages that were used in the previous chapters as well as the **glcm** package. The **glcm** package computes image textures derived from grey-level co-occurrence matrices (GLCMs) in R.

3.3.3 *Procedure*

Step 1: Install and load the necessary packages or libraries

First, install the **glcm** package that will be used in this exercise.

```
> install.packages("glcm")
```

Load the following packages that will be used in this tutorial exercise.

```
> require(sp)
> require(rgdal)
> require(raster)
> require(glcm)
```

Step 2: Load the Landsat stack

Next, set up your working directory.

```
> setwd("E:\\path\\to\\Data\\folder")
```

As we have done before in the previous chapters, create a list of raster layers or bands that will be imported.

```
> rlist<-list.files(getwd(),pattern="img$",
         full.names=TRUE)
```

Next, combine or stack the raster layers.

```
> costz22_6_84 <- stack(rlist)
```

Step 3: Check the attributes

Check the attributes and dimensions of the Landsat 5 TM imagery.

```
> costz22_6_84
```

Step 4: Display the Landsat imagery

Display the composite Landsat 5 TM subset.

```
> plotRGB(costz22_6_84, r=5, g=4, b=3, stretch="lin")
```

Step 5: Compute texture index for band 4

Compute the second order GLCM indices (mean, variance, homogeneity and entropy) for Landsat 5 TM band 4 based on a 3 × 3 window size.

```
> glcm_22_6_84.4 <- glcm(raster(costz22_6_84, layer=4),
  window = c(3, 3), statistics = c("mean", "variance", "
homogeneity", "entropy"))
```

Display glcm texture for band 4, glcm_22_6_84.4 (Fig. 3.2).

```
> plot(glcm_22_6_84.4)
```

It is worth noting that texture can be computed using different approaches depending on the spatial resolution of the remotely-sensed imagery. While texture has been reported to improve land use/cover accuracy (Tso and Mather 2001), it is very difficult to interpret. This is because texture analysis depends on the spatial resolution of the satellite sensor and the size of the homogenous area being classified (Tso and Mather 2001). As a result, we do not intend to analyse the texture images. Rather, the intention is to use texture images in order to test if they can improve land use/cover classification (in the later chapters). While some terms will be unfamiliar to the reader, I encourage the reader to consult the relevant literature listed in the reference for more in depth discussion.

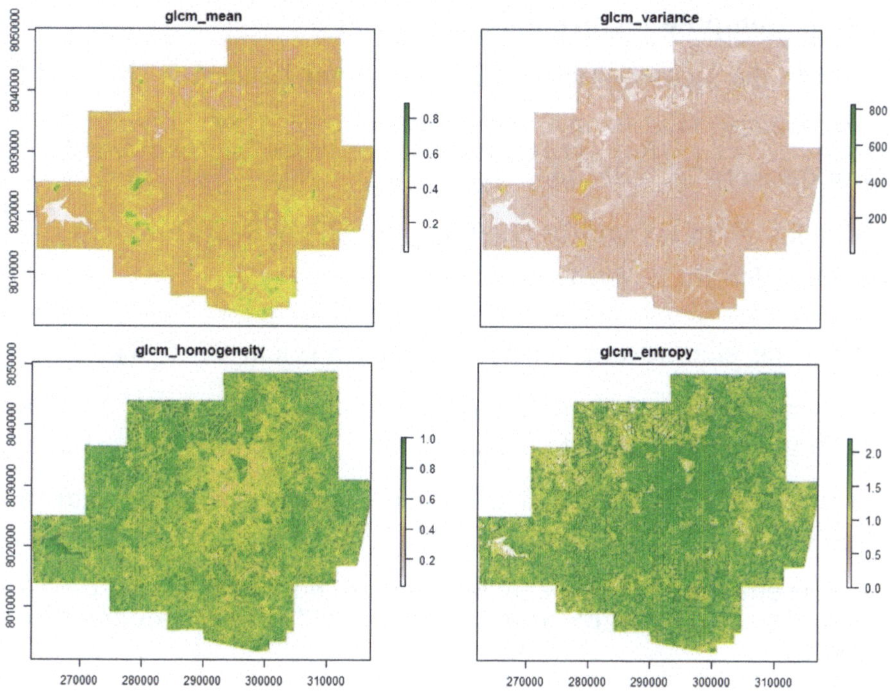

Fig. 3.2 Landsat 5 TM band 4 texture image for the test site

Save the glcm texture images for band 4 (glcm_22_6_84.4). We are going to use the texture images in Chap. 5.

```
>writeRaster(glcm_22_6_84.4$glcm_mean,"G~\\mean_22_6_84
_b4.img", datatype='FLT4S', overwrite = TRUE)

>writeRaster(glcm_22_6_84.4$glcm_variance,"G~\\variance
_22_6_84_b4.img", datatype='FLT4S', overwrite = TRUE)

 >writeRaster(glcm_22_6_84.4$glcm_homogeneity,"G~\\homo
geneity_22_6_84_b4.img", datatype='FLT4S', overwrite =
TRUE)

> writeRaster(glcm_22_6_84.4$glcm_entropy, "G~\\entropy
_22_6_84_b4.img", datatype='FLT4S', overwrite = TRUE)
```

Step 6: Compute texture index for band 5

Compute the second order GLCM indices for Landsat 5 TM band 5.

```
> glcm_22_6_84.5 <- glcm(raster(costz22_6_84, layer=5),
window = c(3, 3), statistics = c("mean", "variance",
"homogeneity", "entropy"))
```

Display glcm texture for band 5, glcm_22_6_84.5 (Fig. 3.3).

```
> plot(glcm_22_6_84.5)
```

Save the glcm texture images for band 5 (glcm_22_6_84.5).

```
writeRaster(glcm_22_6_84.5$glcm_mean, "G~\\mean_22_6_84
_b5.img", datatype='FLT4S', overwrite = TRUE)

> writeRaster(glcm_22_6_84.5$glcm_variance, "G~\\varian
ce_22_6_84_b5.img", datatype='FLT4S', overwrite = TRUE)

 > writeRaster(glcm_22_6_84.5$glcm_homogeneity, "G~\\ho
mogeneity_22_6_84_b5.img", datatype='FLT4S', overwrite
= TRUE)

> writeRaster(glcm_22_6_84.5$glcm_entropy, "G~\\entropy
_22_6_84_b5.img", datatype='FLT4S', overwrite = TRUE)
```

Step 7: Compute texture index for band 7

Compute the second order GLCM indices for Landsat 5 TM band 7.

```
> glcm_22_6_84.6 <- glcm(raster(costz22_6_84, layer=6),
window = c(3, 3), statistics = c("mean", "variance",
"homogeneity", "entropy"))
```

Display glcm texture for band 7, glcm_22_6_84.7 (Fig. 3.4).

```
> plot(glcm_22_6_84.7)
```

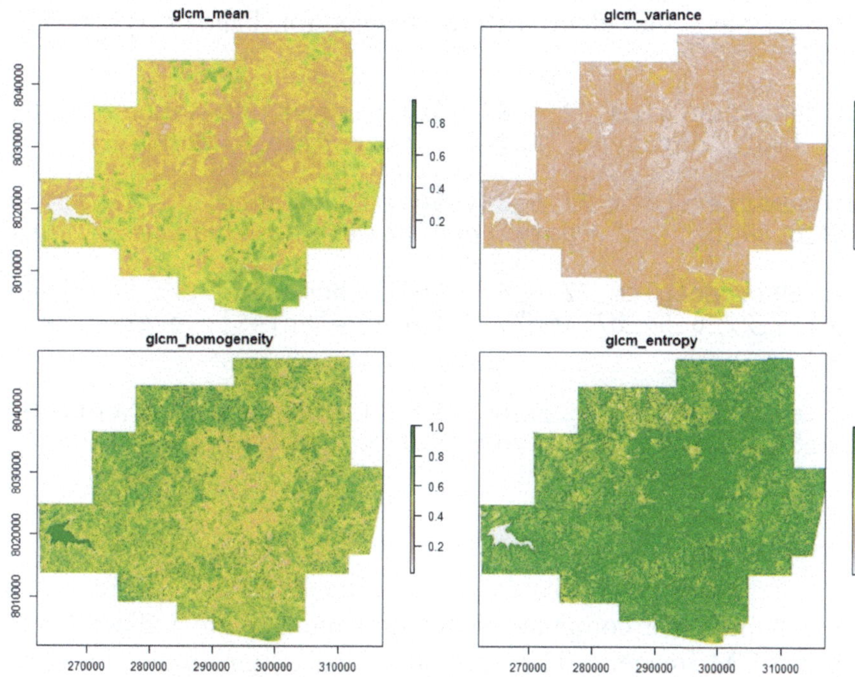

Fig. 3.3 Landsat 5 TM band 5 texture image for the test site

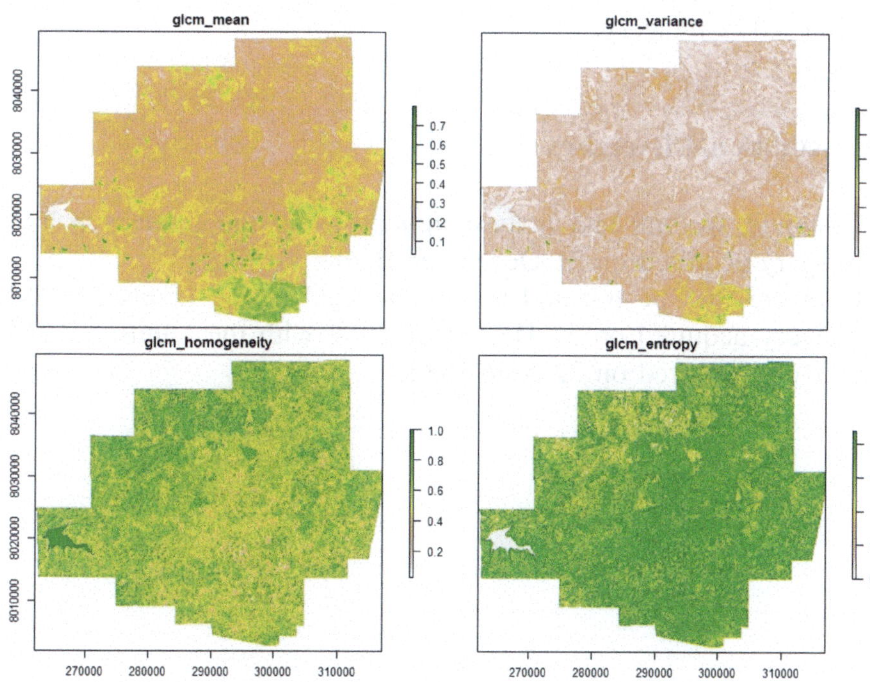

Fig. 3.4 Landsat 5 TM band 7 texture image for the test site

Finally, Save the glcm texture images for band 7 (glcm_22_6_84.7).

```
> writeRaster(glcm_22_6_84.7$glcm_mean, "G~\\mean_22_6_
84_b7.img", datatype='FLT4S', overwrite = TRUE)

> writeRaster(glcm_22_6_84.7$glcm_variance, "G~\\varian
ce_22_6_84_b7.img", datatype='FLT4S', overwrite = TRUE)

> writeRaster(glcm_22_6_84.7$glcm_homogeneity, "G~\\hom
ogeneity_22_6_84_b7.img", datatype='FLT4S', overwrite =
 TRUE)

> writeRaster(glcm_22_6_84.7$glcm_entropy, "G~\\entropy
_22_6_84_b7.img", datatype='FLT4S', overwrite = TRUE)
```

3.4 Summary

In this chapter, we computed vegetation and texture indices using Landsat 5 TM imagery. The vegetation and texture indices will be used for image classification in Chap. 5. In the next chapter, you will learn about image classification using single date and multidate Landsat 5 TM imagery.

3.5 Additional Exercises

i. Compute spectral and texture indices using the Landsat 5 TM imagery acquired on 31 December 1984.
ii. Compare the spectral and texture indices for the Landsat 5 TM imagery acquired on 31 December 1984 with the Landsat 5 TM imagery acquired on 22 June 1984.

References

Berberoglu S, Lloyd CD, Atkinson PM, Curran PJ (2000) The integration of spectral and textural information using neural networks for land cover mapping in the Mediterranean. Comput Geosci 26:385–396

Haralick RM, Shanmugam K, Dinstein I (1973) Textural features for image classification. IEEE Trans Syst Man Cybern SMC-3(6):610–621

Herold M, Goldstein NC, Clarke KC (2003) The spatiotemporal form of urban growth: measurement, analysis and modelling. Remote Sens Environ 86:286–302

Huete AR (1988) Soil-adjusted vegetation index (SAVI). Remote Sens Environ 25:295–309

Leutner B, Horning N (2016) Package 'RStoolbox'. http://bleutner.github.io/RStoolbox. Accessed on 2 April 2017

Mather PM (1999) Computer processing of remotely-sensed images-an introduction. Wiley, New York

Qi J, Chehbouni A, Huete AR, Kerr YH, Sorooshian S (1994) A modified soil adjusted vegetation index. Remote Sens Environ 48:119–126

Rouse JW Jr, Haas R, Schell J, Deering, DW (1974) Monitoring vegetation systems in the great plains with ERTS. In: Third ERTS symposium, NASA, 351, pp 309–317

Shaban MA, Dikshit O (2001) Improvement of classification in urban areas by the use of textural features: the case study of Lucknow City, Uttar Pradesh. Int J Remote Sens 22(4):565–593

Tso B, Mather PM (2001) Classification methods for remotely-sensed data. Taylor and Francis, New York

Wegmann M, Leutner B, Dech S (2016) Remote sensing and GIS for ecologists: using open source software. Pelagic Publishing, Exeter, UK

Zvoleff A (2016) Package 'glcm'. https://cran.r-project.org/web/packages/glcm/glcm.pdf. Accessed on 22 June 2017

Chapter 4
Image Classification

Abstract Image classification refers to the procedure of assigning each pixel in an image to a particular land use/cover class of interest. The purpose of this chapter is to perform supervised classification using single date and multidate Landsat 5 TM imagery and machine learning methods. Five machine learning methods such as k-Nearest Neighbors (KNN), Artificial Neural Networks (ANN), single Decision Trees (DT), Support Vector Machines (SVM) and Random Forests (RF) machine learning classifiers will be used for image classification. The tutorial exercises show that multidate Landsat 5 imagery and the RF method provide relatively good results.

Keywords Supervised classification · k-nearest neighbors · Artificial neural networks · Decision trees · Support vector machines · Random forests

4.1 Overview of Image Classification

Image classification is a procedure for assigning each pixel in an image to a particular class or category based on statistical characteristics of brightness values. In this chapter, we will perform supervised image classification using the COST-corrected single date and multidate Landsat 5 TM imagery.

Electronic supplementary material The online version of this chapter (https://doi.org/10.1007/978-981-13-8012-9_4) contains supplementary material, which is available to authorized users.

C. Kamusoko, *Remote Sensing Image Classification in R*, Springer Geography,
https://doi.org/10.1007/978-981-13-8012-9_4

Generally, image classification is comprised of learning (training), and classification steps. During the first step (that is, learning or training phase), a classification algorithm builds a model by learning from a labeled land use/cover training data set (that is each land use/cover class with associated attributes such as Landsat 5 TM reflectance values). In the second step, the model is used for prediction. More importantly, classification accuracy is also performed using an independent test set.

In this workbook, k-Nearest Neighbors (KNN), Artificial Neural Networks (ANN), single Decision Trees (DT), Support Vector Machines (SVM) and Random Forest (RF) machine learning classifiers will be used for image classification. These are some of the most common classifiers, which are used for remote sensing image processing and classification. Note that, an in-depth review of machine learning classifiers for remote sensing image analysis and classification is beyond the scope of this workbook. Rather, the aim is to provide a brief review with a focus on practical supervised machine learning for remote sensing image classification. Below is a brief summary of the selected machine learning classifiers.

4.1.1 k-Nearest Neighbors (KNN)

The KNN classifier is a non-parametric method used for either classification or regression. The KNN classifier predicts or classifies a new data sample using k-closest sample from the training data. Therefore, the KNN classifier depends largely on how distance between samples is defined. While there are many distance metrics, the Euclidean distance (that is, the straight-line distance between two samples) is commonly used. Since distance between samples is critical, it is recommended to pre-process (centering and scaling) the predictor variables before running the KNN classifier. This removes biases and allows all predictors to be treated equally when computing distance (Kuhn and Johnson 2016). The KNN method is also known as a *non-generalizing* machine learning method, since it simply "remembers" all of its training data. The advantages of the KNN classifier are that: (i) it is a simple classifier that can be easily implemented, and (ii) is appropriate for handling multi-modal classes. However, the KNN classifier requires distance computation of k-nearest neighbors, which can be computationally intensive, especially if the training data set is big.

4.1.2 Artificial Neural Networks (ANN)

The ANN classifiers are mathematical models that mimic the functionality of the human brain for knowledge acquisition, recall, synthesis and problem solving (Yang 2009). Examples of ANN models are: multi-layer perceptron (MPL); Kohonen's Self-organising feature maps (SOM); Hopfield networks; Carpenter/Grossberg classifier; and adaptive resonance theory (ART) model (Tso and Mather 2001). The MLP using the back-propagation learning algorithm is one of the commonly used neural network model (Berberoglu et al. 2000; Tso and Mather 2001). Generally, the MPL consist of three or more layers namely, an input layer, one or more hidden layers and an output layer. The input layer consists of one or more processing elements that present the training data, while the output layer consists of one or more processing elements that store the results of the network. The hidden layer is responsible for the internal representation of data as well as information transformation between input and output layers. The design of the network structure is critical because a small number of neurons in the hidden layer may contain insufficient degrees of freedom to form a representation, while too many neurons may result in overtraining (Mather 1999). The advantages of the ANN classifiers are that they: (i) are free from the normal distribution assumption, (ii) have the ability to generate non-linear decision boundaries, and (iii) the ability to learn complex patterns (Kavzoglu and Mather 2003; Tso and Mather 2001). However, ANN classifiers are prone to overfitting. Furthermore, it is difficult to design an effective neural network.

4.1.3 Single Decision Trees (DT)

The single DT are non-parametric and hierarchical (top-down) splitting classifiers, which use a sequence of decisions that are trained on data for classification and regression problems (Tso and Mather 2001; Miao et al. 2012). Generally, single DT classifiers are composed of a root node, a set of interior nodes and terminal nodes called "leaves" (Tso and Mather 2001). For example, a multispectral remote sensing data set is subdivided into categories based on a splitting mechanism, which chooses the best feature to split the data set (Mather and Koch 2011). The single DT classifier then construct a model of decisions based on actual values of attributes in the data. Some of the most popular DT algorithms are Classification and Regression Tree (CART) (Breiman et al. 1984), Iterative Dichotomiser 3 (ID3) (Quinlan 1986), C4.5 (Quinlan 1993) and C5.0. In this workbook, we will focus on CART (classification and regression trees), which is the one of the most commonly-used DT (Breiman 1993; Quinlan 1993). CART is a binary classifier that uses the Gini impurity index to measure the impurity of a data partition (Miao et al. 2012). The advantages of the single DT classifiers are: (i) they can easily integrate numerical and categorical data; (ii) they require less training time compared to ANN and SVM classifiers (Pal and Mather 2003), and (iii) are free from normal distribution assumptions (Sesnie et al. 2008). However, single DT requires large training samples for classification, and the stability of trees is affected by outliers or small changes in training data (Sesnie et al. 2008).

4.1.4 Support Vector Machines (SVM)

The SVMs are machine learning classifiers based on statistical models first developed by Vladimir Vapnik in the mid-1960s (Kuhn and Johnson 2016). In general, the goal of the SVM model is to locate an optimal decision (separation) boundary that maximizes the margin between two classes. The location of the decision boundary depends only on a subset of training data points that are closest to it. This subset of training data points closest to the decision boundary are known as support vectors (Kuhn and Johnson 2016). Later on improvements were made on the original SVM classifier. Cortes and Vapnik (1995) developed a cost function, while Boser et al. (1992) developed kernel functions in order to allow nonlinear class boundaries. The cost function was developed to quantify misclassification errors or the penalty of having a training point on the wrong side of the decision boundary. The kernel functions such as polynomial, radial basis and hyperbolic tangent were developed to transform training dataset into higher dimensional feature space for nonlinear classification problems (Zhang and Ma 2008; Pal and Mather 2005; Hsu et al. 2010). In this workbook, we use the radial basis SVM classifier. The advantages of SVMs are that: (i) they free from the normal distribution assumption; (ii) the use of an appropriate kernel function can solve complex problems, and (iii) it scales relatively well to high dimensional data. However, SVMs require more training time, especially if the training dataset has many predictors or features. In addition, it is difficult to understand and interpret the final model.

4.1.5 *Random Forest (RF)*

The RF classifier is an ensemble machine learning method, which uses bootstrap sampling to build many single decision tree models (Breiman 2001; Mellor et al. 2013; Rodriguez-Galiano et al. 2012). A random subset of predictor variables (e.g., Landsat bands) is used to split an observation data into homogenous subsets, which are used to build each decision tree model and a prediction (Walton 2008; Mellor et al. 2013). The single decision tree model predictions are then averaged in order to produce the final labeling (Kuhn and Johnson 2016). The out-of-bag (OOB) sample data are used to evaluate performance, while importance measures are computed based on the proportion between misclassifications and OOB sample (Mather and Koch 2011). This provides an unbiased estimation of the generalization error that is used for feature selection (Rodriguez-Galiano et al. 2012). The RF classifier has been successfully used for remotely-sensed image classification because: (i) they can handle large database (e.g., thousands of input numerical and categorical variables); (ii) are free from normal distribution assumptions and (iii) robust to outliers and noise (Rodriguez-Galiano et al. 2012; Mather and Koch 2011). However, it is not easy to interpret the RF model results. In addition, the RF classifier is biased in favor of predictor variables with many different levels of categories (Kuhn and Johnson 2016). As a result, the variable importance scores from the RF classifier are not reliable if a predictor variable has many different levels of categories (Kuhn and Johnson 2016).

4.2 Tutorial 1: Single Date Image Classification

4.2.1 *Objectives*

- Classify single date Landsat 5 TM imagery using selected machine learning classifiers, and
- Compare land use/cover classifications derived from the different machine learning classifiers.

4.2.2 Overview of Packages

In this chapter, we are going to use **sp, rgdal, raster, ggplot2, reshape, grid, gridExtra, RStoolbox, caret, e1071, randomForest, kknn, rpart, nnet, kernlab, rasterVis, maptree, corrplot, doParallel, NeuralNetTools** packages. Some of the packages such as **sp, rgdal, raster, ggplot2, reshape, grid, gridExtra**, and **RStoolbox** have been described in the previous chapter. Table 4.1 provides a brief description of the newly introduced packages.

Table 4.1 Key packages used

Package name	Description
caret	**caret** refers to **C**lassification **A**nd **RE**gression **T**raining. The package was developed by Kuhn (2008) in order to expedite procedures for training and evaluating machine learning models in R. In addition, caret has many functions for splitting data, pre-processing, feature selection, evaluation, and tuning machine learning algorithms
e1071	This package provides functions for support vector machines, bagged clustering, naive Bayes classifier, fuzzy clustering
kernlab	This package provides functions for SVM
randomForest	This package implements Breiman's RF algorithm (based on Breiman and Cutler's original)
kknn	This provides weighted k-Nearest neighbors for classification, regression and clustering
rpart	This package provides functions for recursive partitioning for classification, regression and survival trees
nnet	This is a package for feed-forward neural networks with a single hidden layer, and for multinomial log-linear models
rasterVis	This package provides methods for enhanced visualization and interaction with raster data
maptree	This package provides functions with example data for graphing, pruning, and mapping models from hierarchical clustering, and classification and regression trees
corrplot	This package is a graphical display of a correlation matrix. The package has seven visualization methods such as "circle", "square", "ellipse", "number", "shade", "color" and "pie"
doParallel	This package provides functions for parallel execution of R code on machines with multiple cores or processors or multiple computers
NeuralNetTools	This package provides visualization and analysis tools to aid in the interpretation of neural network models

4.2.3 *Procedure*

In this exercise, we are going to use single date Landsat imagery for classification based on different machine learning classifiers. Following are the image classification steps.

Step 1: Load the necessary packages or libraries

First, we need to install additional libraries.

```
install.packages("caret")
install.packages("e1071")
install.packages("kernlab")
install.packages("rasterVis")
install.packages("maptree")
install.packages("corplot")
install.packages("doParallel")
```

After completing the installation, load all packages.

```
> library(sp)
> library(rgdal)
> library(raster)
> library(reshape)
> library(grid)
> library(gridExtra)
> library(RStoolbox)
> library(caret)
> library(rasterVis)
> library(corrplot)
> library(doParallel)
> install.packages("NeuralNetTools")
> library(NeuralNetTools)
```

Step 2: Load the raster data

Next, set up the working directory where all data sets are located. The data sets consist of Landsat 5 TM imagery (that is, raster data in "img" format), and reference data points (that is, vector data in shapefile format).

```
> setwd("c:\\path\\to\\Data\\folder")
```

Create a list of raster bands that will be used for classification.

```
> rasList <- list.files(getwd(),pattern="img$",
              full.names=TRUE)
```

Combine or stack the raster layers, which you listed before.

```
> rasVars<- stack(rasList)
```

Once we have loaded and created a raster stack, the next step is to check the attributes of the Landsat 5 TM imagery.

```
> rasVars
class       : RasterStack
dimensions  : 1533, 1832, 2808456, 6  (nrow, ncol, ncell, nlayers)
resolution  : 30, 30  (x, y)
extent      : 262065, 317025, 8002765, 8048755  (xmin, xmax, ymin, ymax)
coord. ref.: +proj=utm +zone=36 +south +ellps=WGS84 +towgs84=0,0,0,-0,-0,-0,0 +units=m +no_
defs
```

Plot the Landsat 5 TM imagery as single bands using ***plot()*** function.

```
> TM5_22_06_84 <- plot(rasVars)
```

Next, display the composite Landsat 5 TM imagery using the ***plotRGB()*** function.

```
> TM5_22_06_84 <- plotRGB(rasVars, r=5, g=4, b=3, stret
ch="lin")
```

Step 3: Load the training points

Load the shapefile that contains the training data points using the ***readOGR()*** function.

```
> ta_data <- readOGR(getwd(), "TA_1984")
OGR data source with driver: ESRI Shapefile
Source: "E:/~/Dataset1_Single", layer: "TA_1984"
with 7769 features
It has 1 fields
```

Check the structure of the training points.

```
> str(ta_data)
Formal class 'SpatialPointsDataFrame' [package "sp"] with 5 slots
  ..@ data       :'data.frame':  7769 obs. of  1 variable:
  .. ..$ Cl1984: Factor w/ 5 levels "Agr","Br","GS",..: 4 4 4 4 4 4 4 4 4 4 ...
```

Step 4: Creating a spectral profile

Understanding spectral reflectance curves for different land use/cover classes at different wavelengths is important. Let's now create a spectral profile of the target land use/cover classes from the Landsat 5 TM imagery. Following are the steps for creating a spectral profile.

First, we need to extract the reflectance values using the Landsat 5 TM bands (rasVars) and training points (ta_data). This will create a data frame called "ta" (training area) that contains reflectance values from the Landsat 5 TM bands. Note, this may take a bit of time depending on your computer memory. Be patient!!!

```
ta<-as.data.frame(extract(rasVars, ta_data))
```

Next, let's check the reflectance values in the first lines of the "ta" data frame.

```
> head(ta)
```

	CostzPr_22_6_84.1	CostzPr_22_6_84.2	CostzPr_22_6_84.3	CostzPr_22_6_84.4	CostzPr_22_6_84.5	CostzPr_22_6_84.6
1	0.11951248	0.11166146	0.1818153	0.2839946	0.3645511	0.3382034
2	0.10933009	0.08299519	0.1551487	0.2337471	0.2760634	0.2652575
3	0.10593596	0.06866205	0.1351488	0.2086234	0.1599232	0.1274710
4	0.08557119	0.06866205	0.1284821	0.1918743	0.1654537	0.1598914
5	0.11611834	0.09732833	0.1818153	0.2839946	0.2760634	0.2328372
6	0.09914771	0.07582863	0.1284821	0.2086234	0.2318195	0.2166270

Let's compute the mean reflectance (mr) values for each land use/cover class and each Landsat 5 TM band. We are going to use the ***aggregate()*** function to aggregate the training areas (ta) and the land use/cover class names contained in "ta_data" object.

```
> mr <- aggregate(ta, list(ta_data$Cl1984), mean, na.rm
          = TRUE)
```

Following that, check the mean reflectance values in the first lines of the "mr" data frame.

```
> head(mr)
```

	Group.1	CostzPr_22_6_84.1	CostzPr_22_6_84.2	CostzPr_22_6_84.3	CostzPr_22_6_84.4	CostzPr_22_6_84.5	CostzPr_22_6_84.6
1	Agr	0.06078482	0.04489925	0.12793764	0.3213960	0.4333530	0.3014192
2	Br	0.07225237	0.06131599	0.14929950	0.3486235	0.4501928	0.3152578
3	GS	0.04525830	0.02130178	0.08248431	0.3211411	0.3153480	0.1875433
4	Set	0.09306033	0.08709097	0.17087908	0.3436264	0.3950387	0.3243925
5	Wt	0.04216878	0.01767205	0.06048239	0.1467562	0.1546001	0.1033399

Next, let's remove the first column so that we can use the actual land use/cover names.

```
> rownames(mr) <- mr[,1]
> mr <- mr[,-1]
> mr
```

	CostzPr_22_6_84.1	CostzPr_22_6_84.2	CostzPr_22_6_84.3	CostzPr_22_6_84.4	CostzPr_22_6_84.5	CostzPr_22_6_84.6
Agr	0.06078482	0.04489925	0.12793764	0.3213960	0.4333530	0.3014192
Br	0.07225237	0.06131599	0.14929950	0.3486235	0.4501928	0.3152578
GS	0.04525830	0.02130178	0.08248431	0.3211411	0.3153480	0.1875433
Set	0.09306033	0.08709097	0.17087908	0.3436264	0.3950387	0.3243925
Wt	0.04216878	0.01767205	0.06048239	0.1467562	0.1546001	0.1033399

Now let's create a vector of the legend that shows the color of the land use/cover classes.

```
> mycolor <- c('yellow','grey','green3','red','blue3')
```

Next, let's transform the data frame to a matrix.

```
> mr <- as.matrix(mr)
```

We can now create a spectral profile graph. In order to do that, we need first to create an empty graph using the command below.

```
>plot(0, ylim=c(0.1, 1.0), xlim = c(1,7), type='n',
      xlab="Bands", ylab="Reflectance")
```

Next, add the different land use/cover classes, and title as shown in the command below.

```
> for (i in 1:nrow(mr)){
   lines(mr[i,], type = "l", lwd = 3, lty = 1,
    col = mycolor[i])}


> title(main="Spectral Profile from Landsat 5 TM", font.
         main = 2)
```

Finally, we add the legend. This will generate the spectral profile graph as shown in Fig. 4.1.

```
> legend("topleft", rownames(mr),
         cex=0.8, col=mycolor, lty=1, lwd=3, bty="n")
```

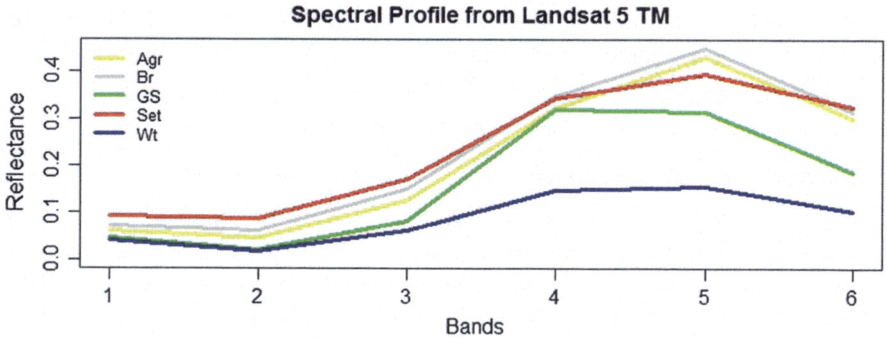

Fig. 4.1 Spectral profile for Landsat 5 TM bands. Note that 6 above refer to Landsat 5 TM band 7

Step 5: Create a data frame with labeled training points containing all reflectance values

In step 4, we used the "ta" data frame to compute mean spectral reflectance and then created a spectral profile. We now need to create a new data frame that contains labeled training points with all reflectance values for each land use/cover class. Use the ***data.frame()*** function to create the "ta_data@data". Note that the "ta_data@data" object has the slots for the land use/cover class names and spectral reflectance values. The @ is a special operator for accessing the objects stored in another object.

```
>ta_data@data=data.frame(ta_data@data,ta[match(rownames
(ta_data@data), rownames(ta)),])
```

After that, let's take a look at the structure of the whole training data set using ***str()*** function. This gives us an overview of the data set.

```
> str(ta_data@data)
```

```
'data.frame':      7769 obs. of  7 variables:
 $ Cl1984             : Factor w/ 5 levels "Agr","Br","GS",..: 4 4 4 4
 4 4 4 4 4 4 ...
 $ CostzPr_22_6_84.1: num  0.1195 0.1093 0.1059 0.0856 0.1161 ...
 $ CostzPr_22_6_84.2: num  0.1117 0.083 0.0687 0.0687 0.0973 ...
 $ CostzPr_22_6_84.3: num  0.182 0.155 0.135 0.128 0.182 ...
 $ CostzPr_22_6_84.4: num  0.284 0.234 0.209 0.192 0.284 ...
 $ CostzPr_22_6_84.5: num  0.365 0.276 0.16 0.165 0.276 ...
 $ CostzPr_22_6_84.6: num  0.338 0.265 0.127 0.16 0.233 ...
```

The data frame consists of 7,769 training points. There are seven variables that comprise a response variable and six predictor variables.

The response variable consists of five target classes (agriculture, bareland, green spaces, settlement and water), while the predictor variables consist of six Landsat 5 TM bands (1, 2, 3, 4, 5 and 7).

Next, check the summary statistics to see if there are NAs.

```
> summary(ta_data@data)
```

```
Cl1984    CostzPr_22_6_84.1 CostzPr_22_6_84.2 CostzPr_22_6_84.3 CostzPr_22_6_84.4 CostzPr_22_6_84.5 CostzPr_22_6_84.6
Agr:2467  Min.   :0.01769   Min.   :0.00000   Min.   :0.008483  Min.   :0.03276   Min.   :0.00507   Min.   :0.0000
Br:3169   1stQu.:0.05163    1stQu.:0.02566    1stQu.:0.101816   1stQu.:0.29237    1stQu.:0.35349    1stQu.:0.2328
GS : 668  Median:0.06181    Median:0.04716    Median:0.135149   Median:0.33424    Median:0.41986    Median:0.2896
Set:1385  Mean   :0.06969   Mean   :0.05681   Mean   :0.139687  Mean   :0.33462   Mean   :0.42028   Mean   :0.2993
Wt :  80  3rdQu.:0.08218    3rdQu.:0.07583    3rdQu.:0.168482   3rdQu.:0.37611    3rdQu.:0.49175    3rdQu.:0.3544
          Max.   0.36728    Max.   :0.39832   Max.   :0.541814  Max.   :0.85347   Max.   :0.82358   Max.   :0.7516
          NA's   :27        NA's   :27        NA's   :27        NA's   :27        NA's   :27        NA's   :27
```

The statistics show that the predictor variables (Landsat 5 TM bands) have missing values (NAs). It is important to treat missing values prior to classification because some machine learning classifiers cannot handle missing values. Next, let's remove the NAs before we proceed.

```
> ta_data@data <- na.omit(ta_data@data)
```

Use the ***complete.cases()*** function to check if the NAs have been removed. Alternatively, you can use ***anyNA()*** function in order to check whether the data contains missing values or not.

```
> Complete.cases(ta_data@data)
```

Step 6: Prepare training and test data sets

Generally, it is recommended to randomly partition the training data set into training, validation and testing sets. In this workbook, we will partition the training data set into the training and testing sets. First, we need to set a pre-defined value using ***set.seed()*** so that the results are repeatable. Setting the seed to the same number will ensure that we get the same result. Choose any number you like to set the seed.

```
> hre_seed<- 27
> set.seed(hre_seed)
```

Next, split the training data set into training and test sets using the ***createDataPartition()*** function from the caret package. In this workbook, 80% of the training data set will be used as a training set, while 20% will be used as a test set. In particular, the training set will be used to find the optimal model parameters as well as to check initial model performance based on repeated cross-validation. The test set will be used for final accuracy assessment.

```
> inTraining <- createDataPartition(ta_data@data$Cl1984,
                   p = .80, list = FALSE)
> training <- ta_data@data[ inTraining,]
> testing  <- ta_data@data[-inTraining,]
```

Notes Note that the response (target) variable is *ta_data@data $Cl1984*, which contains the different land use/cover classes. The "p" parameter shows the percentage of the split, that is we are using p = 0.8. This simply means that data split has 80% training set and 20% test set. The "list" parameter indicates whether to return a list or matrix. In this case, we pass FALSE because we are not returning a list.

Step 7: Check the summary statistics (training and test data sets)

Checking the summary statistics of the training data set and graphical visualization prior to training machine learning classifiers or performing classification is very important. Unfortunately this step is generally ignored because many remote sensing analysts assume that remote sensing imagery and reference (training and testing) data sets derived from ground checks or high resolution imagery are accurate. However, this is not true as error (noise) is introduced during the remote sensing data acquisition or during other important phases such as compilation of reference data sets. Therefore, it is prudent to check the basic univariate descriptive and multivariate statistics as well as check graphical visualization in order to identify unusual anomalies, outliers, distributions, and between predictor correlations.

We start by checking the summary statistics of the training and testing data set using the ***summary()*** function as shown in the commands below.

```
> summary(training)
Cl1984     CostzPr_22_6_84.1 CostzPr_22_6_84.2 CostzPr_22_6_84.3  CostzPr_22_6_84.4 CostzPr
_22_6_84.5 CostzPr_22_6_84.6
 Agr:1969   Min.   :0.01769   Min.   :0.00000   Min.   :0.008483   Min.   :0.03276   Min.
 :0.00507   Min.   :0.0000
 Br :2519   1st Qu.:0.05163   1st Qu.:0.02566   1st Qu.:0.101816   1st Qu.:0.29237   1st Q
u.:0.35349   1st Qu.:0.2328
 GS : 535   Median :0.06181   Median :0.04716   Median :0.135149   Median :0.33424   Median
 :0.41986   Median :0.2896
 Set:1108   Mean   :0.06989   Mean   :0.05706   Mean   :0.139997   Mean   :0.33493   Mean
 :0.42096   Mean   :0.2998
 Wt :  64   3rd Qu.:0.08218   3rd Qu.:0.07583   3rd Qu.:0.168482   3rd Qu.:0.37611   3rd Q
u.:0.49175   3rd Qu.:0.3544
            Max.   :0.36728   Max.   :0.39832   Max.   :0.541814   Max.   :0.85347   Max.
 :0.82358   Max.   :0.7516

> summary(testing)
Cl1984     CostzPr_22_6_84.1 CostzPr_22_6_84.2 CostzPr_22_6_84.3 CostzPr_22_6_84.4 CostzPr_2
2_6_84.5 CostzPr_22_6_84.6
 Agr:492   Min.   :0.02108   Min.   :0.00000   Min.   :0.01515   Min.   :0.03276   Min.   :
0.0106    Min.   :0.005895
 Br :629   1st Qu.:0.05163   1st Qu.:0.02566   1st Qu.:0.10182   1st Qu.:0.29237   1st Qu.:
0.3590    1st Qu.:0.232837
 GS :133   Median :0.06181   Median :0.04716   Median :0.13515   Median :0.33424   Median :
0.4199    Median :0.297678
 Set:277   Mean   :0.06887   Mean   :0.05578   Mean   :0.13845   Mean   :0.33338   Mean   :
0.4176    Mean   :0.297316
 Wt : 16   3rd Qu.:0.08218   3rd Qu.:0.07583   3rd Qu.:0.16848   3rd Qu.:0.36774   3rd Qu.:
0.4890    3rd Qu.:0.354414
           Max.   :0.21794   Max.   :0.21199   Max.   :0.33515   Max.   :0.83672   Max.   :
0.7351    Max.   :0.64619
```

The output above for both the training and testing data sets shows the number of land use/cover training points (Cl1984) in the first column, which is represented by Agr (agriculture), Br (bareland), green space (GS), settlement (Set), and water (Wt). Columns two to seven provides descriptive statistics for each band (note that "CostzPr_22_6_84.6" actually represents Landsat 5 TM band 7). While the output shows useful information such as the highly imbalanced nature of land use/cover classes, it is difficult to understand nature of distribution (normal or skewed), or to detect other problems such as outliers and collinearity.

Step 8: Visualization

Next, we further explore training data using graphical visualizations. Graphs provide a useful way of understanding the training data set. In addition, graphs can help the remote sensing analysts to select the most appropriate methods for pre-processing, transformation and classification. Here, we are going to use the ***featurePlot()*** function (which is a wrapper for different lattice plots) and the ***ggplot2()*** function in order to visualize the training data set. In addition, we are going to use the ***corrplot()*** function to create band correlation matrices.

Let's start by creating density estimation plots (density plots) for each attribute by class value using the ***featurePlot()*** function. The density plots summarize the distribution of the data. This highlights useful structures like linear separability of attribute values into classes (Fig. 4.2).

```
> featurePlot(x = training[, 2:7],
        y = training$Cl1984,
        plot = "density",
        labels=c("Reflectance", "Density distribution"),
              scales = list(x = list(relation="free"),
                            y = list(relation="free")),
        layout = c(3, 2),
        auto.key = list(columns = 3))
```

Clearly, the density plots show that bands 1 and 2 (that is "CostzPr_22_6_84.1" and "CostzPr_22_6_84.2") display significant skewness. We can also calculate skewness of the bands using the ***skewness()*** function available in the **e1071** package.

```
> skewnessValues <- apply(training[, 2:7], 2, skewness)
> skewnessValues

CostzPr_22_6_84.1  CostzPr_22_6_84.2  CostzPr_22_6_84.3
1.837373             1.3659797          0.8632132

CostzPr_22_6_84.4  CostzPr_22_6_84.5  CostzPr_22_6_84.6
 0.5591144           0.2176414         0.4032507
```

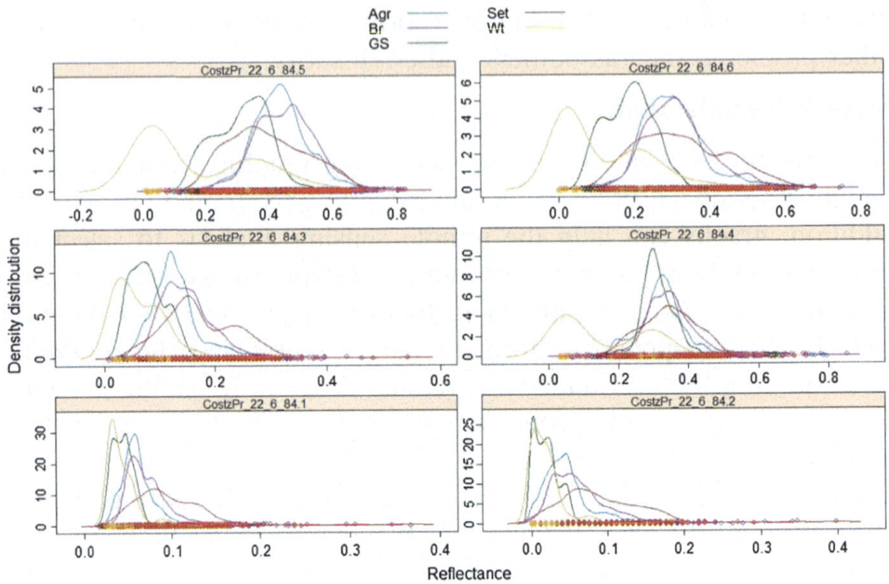

Fig. 4.2 Density plots for the training data set

Generally, the distribution is symmetric when the skewness value is zero or closer to zero. The output shows that the skewness values for bands 1 and 2 are 1.8 and 1.4, respectively. This indicates that the distributions are indeed skewed.

Figure 4.2 also shows that some land use/covers classes such as water in bands 5 and 6 are bimodal. In addition, the density plots suggest presence of outliers. Therefore, we need to check for outliers.

Next, let's use the ***featurePlot()*** function to display the box and whisker plots for each attribute by class value. This summarizes the spread of attributes. Note that box plots summarize the distribution of a given attribute by showing a box for the 25th and 75th percentile, a line in the box for the 50th percentile (median) and a dot for the mean. The whiskers show 1.5 times the height of the box (i.e., Inter Quartile Range), which indicates the expected range of the data. Data beyond those whiskers is assumed to be an outlier and marked with a dot (Fig. 4.3).

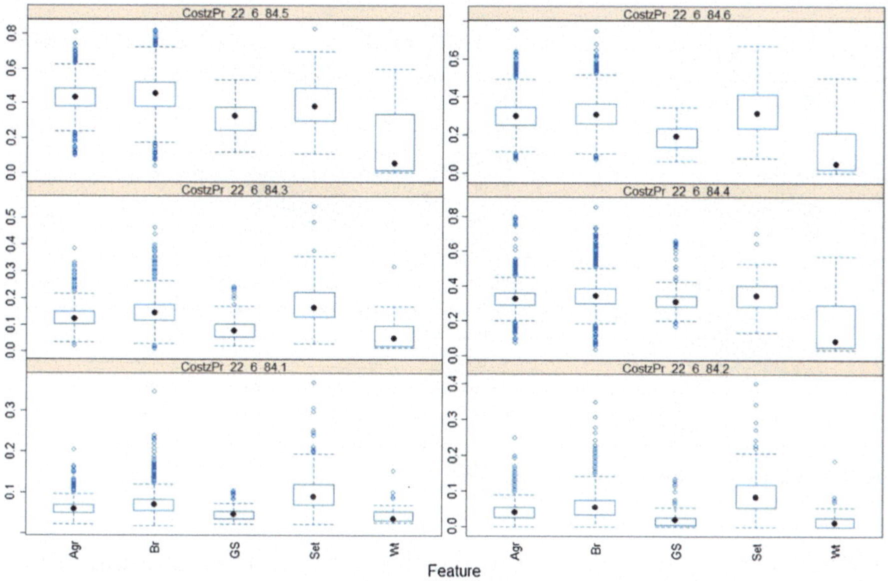

Fig. 4.3 Box plots for the training data set

```
> featurePlot(x=training[,2:7],y=training$Cl1984,
              plot="box",
              scales=list(y=list(relation="free"),
              x=list(rot=90)),
              layout=c(2,3),
              auto.key=list(colmns=2))
```

Figure 4.3 shows the presence of outliers (that is extreme reflectance values), especially for the bareland and agriculture classes. This is problematic for some machine learning classifiers. One solution would be to remove the outliers from the training data sets. However, this could result in throwing away training data, which normally is expensive and time consuming to generate. Therefore, it is better to understand the reason for having outliers in the first place. While the presence of some outliers can be attributed to human error during compilation of reference data, in this case the presence of outliers in the agriculture and bareland is partly attributed to soil variability in the test site. Figure 4.1 indicated almost similar spectral reflectance between the agriculture and bareland because both classes are

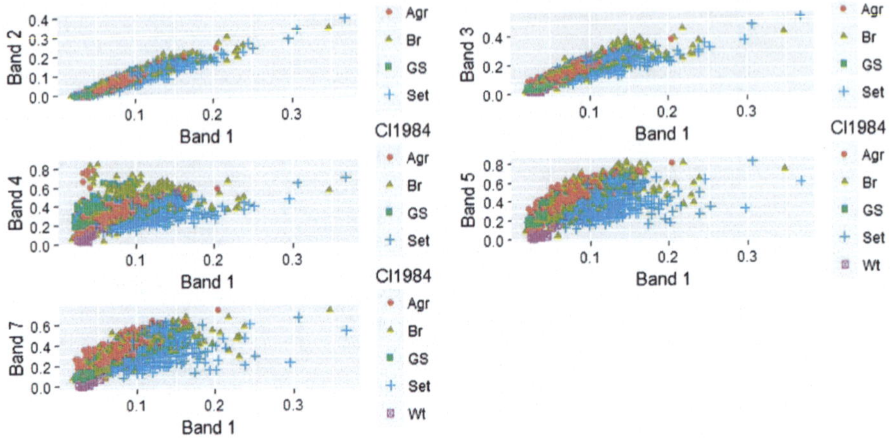

Fig. 4.4 Scatter plots for the training data set

dominated by bare soils during the dry season. This example, illustrate the importance of understanding the geographical characteristics of the test site.

Let's also create scatter plots in order to check the relationships between two bands in the feature space using ***ggplot()*** function.

```
> Band1_2<- ggplot(data = ta_data@data, aes(CostzPr_22_
6_84.1, CostzPr_22_6_84.2)) + geom_point(aes(shape= Cl1
984, colour= Cl1984)) + labs(x="Band 1", y = "Band 2")
> Band1_3<-ggplot(data = ta_data@data, aes(CostzPr_22_6
_84.1, CostzPr_22_6_84.3)) +  geom_point(aes(shape= Cl1
984, colour= Cl1984)) + labs(x="Band 1", y = "Band 3")
> Band1_4<-ggplot(data = ta_data@data, aes(CostzPr_22_6
_84.1, CostzPr_22_6_84.4)) +  geom_point(aes(shape= Cl1
984, colour= Cl1984)) + labs(x="Band 1", y = "Band 4")
> Band1_5<- ggplot(data = ta_data@data, aes(CostzPr_22_
6_84.1, CostzPr_22_6_84.5)) + geom_point(aes(shape= Cl1
984, colour= Cl1984)) + labs(x="Band 1", y = "Band 5")
> Band1_7<-ggplot(data = ta_data@data, aes(CostzPr_22_6
_84.1, CostzPr_22_6_84.6)) +  geom_point(aes(shape= Cl1
984, colour= Cl1984)) + labs(x="Band 1", y = "Band 7")
```

To display the scatter plots, use ***grid.arrange()*** so that all the plots appear on the same sheet (Fig. 4.4).

```
> grid.arrange(Band1_2, Band1_3, Band1_4, Band1_5, Band
1_7, ncol=2)
```

Figure 4.4 indicates that bands 1 and 2 as well as bands 1 and 3 have difficulty to separate land use/cover classes. In addition, Fig. 4.4 suggests that there is high between band (predictor) correlations, especially for bands 1 and 2, and bands 1 and 3. Therefore, it is better to check between band correlations. So let's check correlation for the six Landsat 5 TM bands using the ***cor()*** and ***corrplot()*** functions.

```
> bandCorrelations <- cor(training[, 2:7])
```

Let's check the dimension of the object "bandCorrelations".

```
> dim(bandCorrelations)
```

```
[1] 6 6
```

Next, check the band correlations.

```
> bandCorrelations
```

	CostzPr_22_6_84.1	CostzPr_22_6_84.2	CostzPr_22_6_84.3	CostzPr_22_6_84.4	CostzPr_22_6_84.5	CostzPr_22_6_84.6
CostzPr_22_6_84.1	1.0000000	0.9610577	0.8966841	0.5365460	0.5901836	0.7354192
CostzPr_22_6_84.2	0.9610577	1.0000000	0.9580730	0.6419002	0.6828810	0.8056505
CostzPr_22_6_84.3	0.8966841	0.9580730	1.0000000	0.6560069	0.7907537	0.8759879
CostzPr_22_6_84.4	0.5365460	0.6419002	0.6560069	1.0000000	0.6714087	0.5553501
CostzPr_22_6_84.5	0.5901836	0.6828810	0.7907537	0.6714087	1.0000000	0.9059390
CostzPr_22_6_84.6	0.7354192	0.8056505	0.8759879	0.5553501	0.9059390	1.0000000

Since it is difficult to read the R output, graphic visualization is better. So let's display the following correlation matrices as numbers, a simple graph or as mixed plots (numbers and plots). This makes it easier to interpret the correlation coefficient. Run the commands below.

```
> corrplot(bandCorrelations, method = "number", type =
"upper")
```

```
> corrplot(bandCorrelations, method = "color", order =
"hclust", type = "lower")
```

```
>corrplot.mixed(bandCorrelations,lower.col="black",
number.cex = .7, upper = "color")
```

Figure 4.5a–c show correlation matrices for the six Landsat 5 TM bands (that is, the predictor variables). Generally, the correlation matrix shows positive correlations in dark blue, while negative correlations are shown in red color. The color intensity and the size of

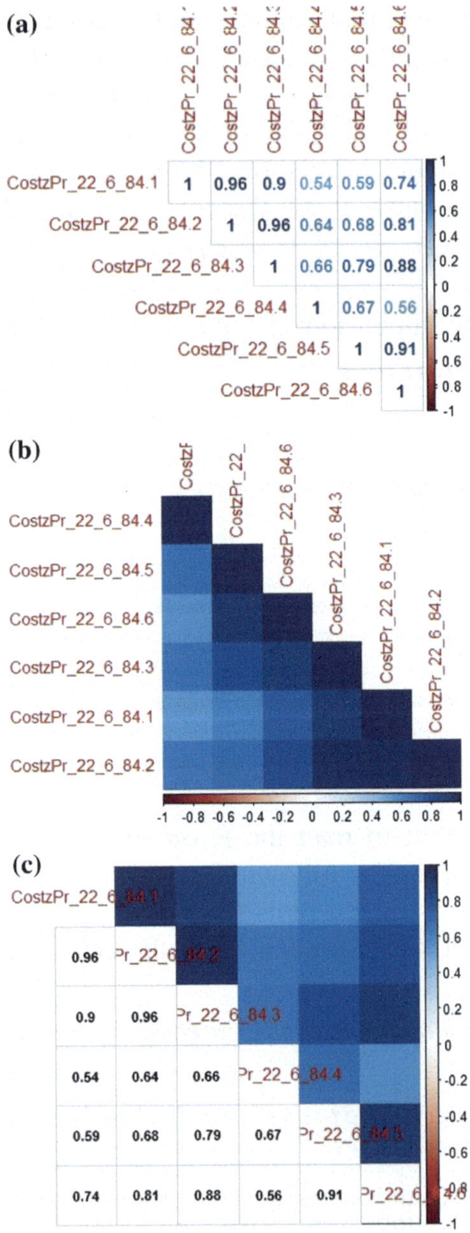

Fig. 4.5 **a** Landsat 5 TM band correlation matrix displayed as numbers. **b** Landsat 5 TM band correlation matrix displayed as a simple graph. **c** Landsat 5 TM band correlation matrix displayed as mixed plots

the square are proportional to the correlation coefficients. Clearly, bands 1 and 2, bands 1 and 3, bands 5 and 7 are highly correlated (correlation coefficient is above 0.9). Note that it is recommended to remove the highly correlated bands or perform a principal component analysis in order to reduce redundancy and improve computation efficiency. However, in this tutorial exercise we are going to use all the bands since we have only six bands.

In summary, the preliminary statistical and graphical analysis shows the training data has skewed and bimodal distributions as well as many outliers. In addition, we observed relatively high class imbalance, and some of the predictors such as bands 1 and 2 are highly correlated. Therefore, it would be inappropriate to use conventional classifiers such as maximum likelihood that depend on the assumption of normal of distribution. Although machine learning classifiers are not affected by the normal distribution assumption, problems such as class imbalance, outliers and between predictor collinearity have negative impact on performance and accuracy. Therefore, it is important to evaluate many machine learning classifiers in order to find the most appropriate one, which improves performance and accuracy. In this workbook, we are going to use the **caret** package to build and evaluate different machine learning classifiers. We will start by setting up tuning parameters, and then move on to train and evaluate each machine learning classifier.

Step 9: Set-up model tuning parameters

Setting up model tuning parameters is important because most classifiers have at least one key parameter, which control the complexity of the model. Failure to select appropriate model parameters may result in problems such as overfitting. In this workbook, we are going to use the ***trainControl()*** function from the **caret** package to evaluate—based on resampling—the effect of model tuning parameters on performance. This will also be used to select the "optimal" model as well as to estimate model performance. So let's set the model tuning parameters as shown below.

```
> fitControl<- trainControl(method = "repeatedcv",
                number = 5,
                repeats = 5)
```

We specify the "method", "number" and "repeats" parameters of the ***trainControl()*** function. The "method" parameter contains resampling methods such as "repeatedcv", "boot", "cv", "LOOCV" etc. In this workbook, we are going to use repeatedcv that is the repeated cross-validation. The "number" parameter refers to the number of folds or number of resampling iterations. The "repeats" parameter provides sets of folds in order to compute repeated cross-validation. In this case, we are running a 5-fold cross validation and repeating it five times (that is, number = 5 and repeats = 5).

Notes Cross-validation is a procedure whereby the training data set is subdivided into a number of mutually exclusive groups. For the k-fold cross-validation, the initial training data set is randomly partitioned into k mutually exclusive subsets or folds, which are approximately of the same size. Training and testing is then performed k times. That is, the k-fold cross-validation repeats the model construction using different subsets of the available training data, and then evaluates the model only on unseen data during model building. Note that I have specified k value of 5 in this workbook because it is computationally efficient. However, the k value is recommended to be 5 or 10 in order to minimize bias (Kuhn and Johnson 2016).

Step 10: Train KNN classifier

The ***train()*** function from the **caret** package is the workhorse for tuning machine leaning classifiers in this workbook. The function sets up a grid of tuning parameters, fits each model, and calculates a resampling based performance measure. For each training data set, the performance of held-out samples is calculated and summary statistics are provided. The final model with the optimal resampling statistic is selected.

For training the KNN classifier, we specify the ***train()*** function as follows. First, we define "cl1984 ~ .," which denotes a formula for using all attributes in our classifier, and "cl1984" is the response (target) variable. The "data" contains the predictor variables (in this case, Landsat 5 TM reflectance values). The "method" is defined as "kknn", while the tune control (trControl) is specified as "fitControl" (that we have previously defined in step 9). All results will be saved in the object "knnFit". Generally, it is recommended to center and scale all predictor variables prior to executing the KNN model in order to reduce bias among the predictor variables during distance computation (Kuhn and Johnson 2016). We will start to train the KNN classifier as shown below.

```
> set.seed(hre_seed)
> knnFit<- train(Cl1984 ~ ., data = training,
                  method = "kknn",
                  preProcess = c("center", "scale"),
                  trControl = fitControl)
Attaching package: 'kknn'
The following object is masked from 'package:caret':
    contr.dummy
```

Use the ***print()*** function to check model results.

```
> print(knnFit)

k-Nearest Neighbors
6195 samples
   6 predictor
   5 classes: 'Agr', 'Br', 'GS', 'Set', 'Wt'
Pre-processing: centered (6), scaled (6)
Resampling: Cross-Validated (5 fold, repeated 5 times)
Summary of sample sizes: 4958, 4956, 4955, 4956, 4955,
4956, ...
Resampling results across tuning parameters:


 kmax  Accuracy   Kappa
  5     0.7033111  0.5737874
  7     0.7030206  0.5733515
  9     0.7026977  0.5725810
Tuning parameter 'distance' was held constant at a val-
ue of 2
Tuning parameter 'kernel' was held constant at a value
of optimal
Accuracy was used to select the optimal model using the
largest value.
The final values used for the model were kmax = 5, dis-
tance = 2 and kernel = optimal.
```

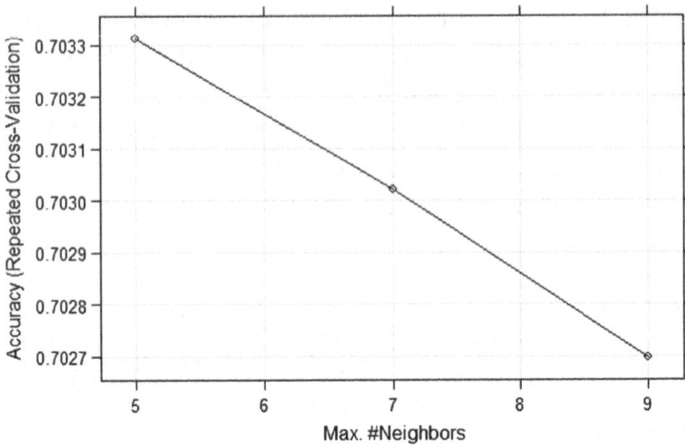

Fig. 4.6 Repeated cross-validation (based on overall accuracy) profile for the KNN model

The output shows that 6,195 training samples were used for training. There are six predictors (that is, six Landsat 5 TM bands) and five land use/cover classes, which are agriculture (Agr), bareland (Br), green space (GS), settlement (Set) and water (Wt). During the KNN model tuning, pre-processing was done. The cross-validation results also show the sample sizes, and final values that were selected for the best model. Note that the ***train()*** method has helped us to:

- cross validate the KNN model;
- tune the parameters for optimal model performance; and
- select the optimal model based on evaluation metrics.

We can display the model training performance based on overall accuracy as shown in Fig. 4.6 using the command below.

```
> plot(knnFit)
```

Figure 4.6 shows the repeated cross-validation profile for the KNN model, which is based on overall accuracy. The optimal number of neighbors is 5. Notice that there is a slight decrease from “kmax = 5” to “kmax = 7”.

Check the parameters of the best model.

```
> knnFit$finalModel

Call:
kknn::train.kknn(formula = .outcome ~ ., data = dat, km
ax = param$kmax,      distance = param$distance, kernel
= as.character(param$kernel))
Type of response variable: nominal
Minimal misclassification: 0.2713479
Best kernel: optimal
Best k: 1
```

Next, let's display variable importance using the ***varImp()*** function.

```
> knn_varImp <- varImp(knnFit, compete = FALSE)
> plot(knn_varImp)
```

Figure 4.7 shows that Landsat 5 TM bands 7, 3, 1, 2 and 5 are the most informative predictor variables for the settlement class, while band 4 has the lowest contribution. One possible explanation is that band 4 exhibits significant spectral confusion between settlement class, and the agriculture and bareland classes (see Fig. 4.1). Note that in the test site, high density urban areas are characterized by gravel (dirt) road, which have the same spectral patterns as bareland and agriculture areas. However, a closer look at Fig. 4.1 also shows that band 7 (CostzPr_22_6_84.6) also exhibit significant spectral confusion between settlement, and the agriculture and bareland classes. Interestingly, for the bareland class, band 1 has the greatest contribution, while bands 4 and 5 have the lowest contribution. Although the analysis of variable importance provide some important insights, it is difficult to understand how the KNN model came up with the variable importance ranking. Therefore, great care should be taken when analyzing the variable importance. It is also important to check if the most informative predictors were selected based on geographical characteristics of the test site and remote sensing principles. For the KNN model, the variable importance ranking is misleading as one would expect band 5 to have more influence or contribution.

Next, let's perform prediction using the ***predict()*** function. We are going to pass two arguments. The first parameter is the trained model ("knnFit"), and the second parameter "newdata" holds the testing data

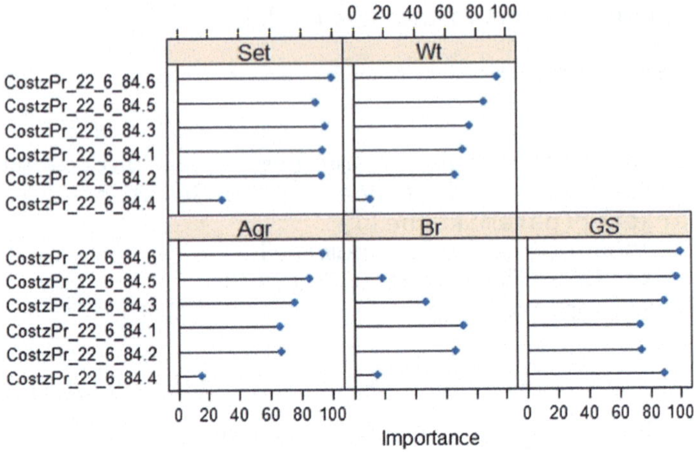

Fig. 4.7 Variable importance for the KNN model

("testing"). The ***predict()*** function will return a list that we are going to save as "pred_knnFit".

```
> pred_knnFit<- predict(knnFit, newdata = testing
```

Before performing accuracy assessment, we need to be familiar with key terms. The confusion matrix (error matrix), which is a cross-tabulation of the reference (observed data) and predicted classes (classified) is the most common method used for land use/cover accuracy assessment (Table 4.2). The diagonal areas represent land use/cover classes that were correctly predicted or classified, while the off-diagonals represent classification errors (that is, errors of commission and omission). The columns represent reference land use/cover (derived from ground checks, high resolution aerial photographs or satellite imagery), while the rows represents the predicted or classified land use/cover class.

Let's consider a case where we have binary land use/cover classification problem (Table 4.2). In this case, we are performing accuracy assessment for agriculture (Agr) and non-agriculture (Non-agr) classes (Table 4.2). True positive (TP) quantifies the number of correctly allocated positives, that is the correct agriculture class. The true negative (TN) quantifies the number of correctly allocated non-agriculture class. The false positive (FP) quantifies the number of non-agriculture pixels classified as agriculture (error of

Table 4.2 A 2 × 2 confusion matrix with notation

	Reference (observed)			Total
Predicted (classified)		Agr	Non-agr	
	Agr	*TP*	FN	A
	Non-agr	FP	*TN*	N
Total		A^1	N^1	

Note
Agr agriculture
Non-agr non-agriculture
TP True positives
FN False negatives (errors of omission)
FP False positives (errors of commission)
TN True negatives
A is the sum of true positives and false negative
N is the sum of false positives and true negative
A^1 is the sum of true positives and false positives
N^1 is the sum of false negatives and true negatives

commission), while the false negative (FN) quantifies the number of agriculture pixels classified as non-agriculture (error of ommission).

You will notice that the confusion matrix produces several accuracy assessment metrics. In this workbook, we will focus on accuracy assessment metrics commonly used in remote sensing (Foody 2010) such as the overall accuracy, producer's accuracy (sensitivity and specificity) and user's accuracy (positive prediction value and negative prediction value). The overall accuracy is the proportion of correctly classified land use/cover samples. The producer's accuracy is the land use/cover map accuracy from the point of view of the map maker (the producer). In other words, the producer's accuracy determines how well the classification model detects a given class. The user's accuracy is the accuracy from the point of view of a map user. That is, the user's accuracy informs us how often the class on the land use/cover map will actually be present on the ground. Note that, we will ignore the kappa coefficient kappa because it is highly correlated with overall accuracy and therefore, redundant (Pontius and Millones 2011; Olofsson et al. 2014). The formulas are expressed as follows:

$$Overall\,accuracy = (TP + TN)/(TP + FN + FP + TN)$$
$$Producer's\,accuracy\,(\text{sensitivity}) = TP/A^1$$
$$Producer's\,accuracy\,(\text{specificity}) = TN/N^1$$
$$User's\,accuracy\,(\text{positive prediction value}) = TP/A$$
$$User's\,accuracy\,(\text{negative prediction value}) = TN/N$$

Next, let's check the accuracy metrics for the KNN model using the ***confusionMatrix()*** function from the **caret** package in R.

```
> confusionMatrix(data = pred_knnFit, testing$Cl1984)
```

Following are the confusion matrix results, overall statistics and class statistics.

```
Confusion Matrix and Statistics
          Reference
Prediction Agr  Br  GS Set  Wt
       Agr 359 112  18  19   1
       Br  107 445  12  39   1
       GS   17  18  93   9   0
       Set   9  52   6 208   1
       Wt    0   2   4   2  13


Overall Statistic
               Accuracy : 0.7227
                 95% CI : (0.6996, 0.7449)
    No Information Rate : 0.4066
    P-Value [Acc > NIR] : <2e-16
                  Kappa : 0.6021
 Mcnemar's Test P-Value : 0.2216
Statistics by Class:
                     Class: Agr Class: Br Class: GS Class: Set Class: Wt
Sensitivity              0.7297    0.7075   0.69925     0.7509  0.812500
Specificity              0.8578    0.8268   0.96888     0.9465  0.994775
Pos Pred Value           0.7053    0.7368   0.67883     0.7536  0.619048
Neg Pred Value           0.8719    0.8049   0.97163     0.9457  0.998034
Prevalence               0.3180    0.4066   0.08597     0.1791  0.010343
Detection Rate           0.2321    0.2877   0.06012     0.1345  0.008403
Detection Prevalence     0.3290    0.3904   0.08856     0.1784  0.013575
Balanced Accuracy        0.7937    0.7671   0.83407     0.8487  0.903637
```

Let's examine the agriculture class as an example. The confusion matrix results show that 359 pixels were correctly classified as agriculture. However, 107 bareland (Br), 17 green space (GS) and 9 settlement (Set) pixels were wrongly classified as agriculture. These 131 misclassified pixels represent the false positives or errors of commission. In contrast, 112 bareland (Br), 18 green space (GS), 19 settlement (Set) and 1 water (Wt) pixels were incorrectly classified as non-agriculture. These 150 misclassified pixels represent the false negatives or errors of omission.

Next, let's take a look at the overall statistic results. The overall accuracy is 72.3%, while the "No Information Rate" is about 41%. The "No Information Rate" is the best guess given no information beyond the overall distribution of the classes that are being classified. Therefore, the KNN classifier is better than the best guess model.

Next, let's examine classification accuracy performance of each land use/cover class. First, we examine the producer's accuracy in terms of sensitivity and specificity. The sensitivity for the agriculture (agr) class is 73%, while the specificity is 86%. This means that the KNN classifier correctly identified the agriculture (Agr) class 73% of the time. However, KNN classifier correctly identified non-agriculture (non-agr) classes (that is bareland, green spaces, settlement and water) 86% of the time. This pattern is also observed for the bareland, green spaces, settlement and water classes. Therefore, the KNN classifier is better at detecting the true negative rate rather than the true positive rate.

Second, we examine the user's accuracy in terms of the positive predicted values (Pos Pred Value) and negative predicted values (Neg Pred Value). The Pos Pred Value for the agriculture (agri) class is 71%, while the Neg Pred Value is 87%. This means that from the map user's perspective, the KNN classifier correctly identified agriculture (Agr) 71% of the time. However, the KNN classifier correctly identified non-agriculture (non-agr) classes (that is bareland, green spaces, settlement and water) 87% of the time. Again, the KNN classifier is also better at classifying the non-agriculture classes.

The accuracy assessment results highlight important insights on land use/cover classification. For example, the confusion matrix reveals significant spectral confusion between agriculture and bareland classes. This is not surprising given the high quantity of outliers observed in Fig. 4.3. Furthermore, the KNN is biased towards high specificity and negative predicted values. As a result, the KNN model is poor at detecting the true positive classes. This information can be used to improve land use/cover classification.

Step 11: Train ANN classifier

Next, let's train the ANN classifier. In R, I use the **nnet** (neural network) package since it is easy to adjust the weight decay. However, the **nnet** package is generally restricted to one hidden layer, which is a limitation. Here, we use "size" and "decay" as the primary model tuning parameters. The parameter "size" refers to the number of nodes in the hidden layer, while "decay" refers to the size of the weight decay. Remember, the ANN model has a tendency to overfit. As a result, a weight decay—which is a penalization method for regularizing or controlling the model—is used to counter overfitting.

For training the ANN model, we specify the ***train()*** method as we have done in step 10. We define the response (target) variable as "cl1984 ~ .," the "data" as training, and tune control ("trControl") as "fitControl". Here, we define the "method" as "nnet" since we are running a neural network classifier. All results will be saved in the object "annFit". We will start to train the ANN classifier as shown below.

```
> set.seed(hre_seed)
> annFit <- train(Cl1984 ~ ., data = training,
                  method = "nnet",
                  preProcess = c("center", "scale"),
                  trControl = fitControl)
```

As I mentioned before, the ANN classifier is a connection of input/output nodes whereby each connection has a weight associated with it. Therefore, the network learns by adjusting the weights in order to predict or identify the correct land use/cover class during training. The following is part of the program output, which shows progress messages that are generated automatically while the ANN is being trained.

```
iter  80 value 5469.241790
iter  90 value 5459.317252
iter 100 value 5452.861744
final  value 5452.861744
stopped after 100 iterations
```

The above output shows that the network converged after 100 iterations. Next, let's use the ***print()*** function to check the model results.

```
> print(annFit)

Neural Network
6195 samples
   6 predictor
   5 classes: 'Agr', 'Br', 'GS', 'Set', 'Wt'

Pre-processing: centered (6), scaled (6)
Resampling: Cross-Validated (5 fold, repeated 5 times)
Summary of sample sizes: 4958, 4956, 4955, 4956, 4955,
4956, ...
Resampling results across tuning parameters:

  size  decay  Accuracy   Kappa
  1     0e+00  0.5220345  0.2758490
  1     1e-04  0.5210339  0.2744500
  1     1e-01  0.5201294  0.2748127
  3     0e+00  0.5880879  0.4006588
  3     1e-04  0.5861496  0.3980396
  3     1e-01  0.5918042  0.4067903
  5     0e+00  0.6095596  0.4326138
  5     1e-04  0.6098188  0.4325456
  5     1e-01  0.6156594  0.4412318
Accuracy was used to select the optimal model using the
largest value.
The final values used for the model were size = 5 and
decay = 0.1.
```

The ANN classifier results show that 6,195 training samples were used for training. Six are predictors (that is, six Landsat 5 TM bands), and five land use/cover classes represent the response (target) variable. In addition, pre-processing was done. The cross-validation results also provide the sample sizes and the final values that were selected for the best model. The “size” parameter specifies the number of hidden processing nodes used in the neural network.

Next, let’s display the model training performance based on overall accuracy as shown in Fig. 4.6 using the command below.

```
> plot(annFit)
```

Figure 4.8 shows the weight decay values that were evaluated with a single hidden layer with sizes ranging between 1 and 5 hidden nodes. The optimal ANN model has 5 hidden nodes and a weight decay of 0.1.

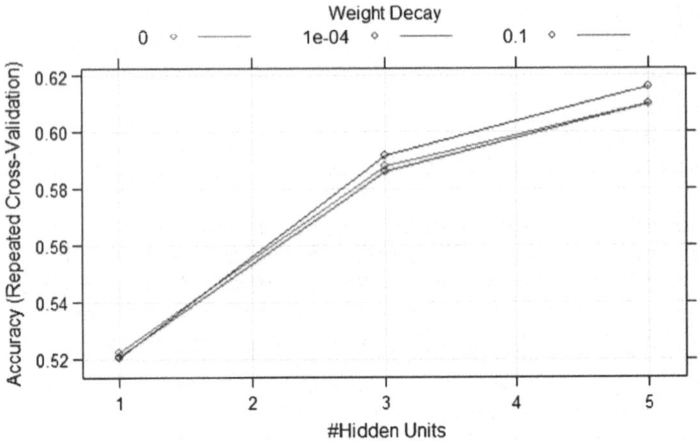

Fig. 4.8 Repeated cross-validation (based on overall accuracy) profile for the ANN model

Next, check the parameters of the best model.

```
> annFit$finalModel
a 6-5-5 network with 65 weights
inputs: CostzPr_22_6_84.1 CostzPr_22_6_84.2 CostzPr_22_
6_84.3 CostzPr_22_6_84.4 CostzPr_22_6_84.5 CostzPr_22_6
_84.6
output(s): .outcome
options were - softmax modelling  decay=0.1
```

We trained a network with 6 variables as input data (that is, the Landsat bands), 5 nodes in the hidden layer and five classes in the output layer. In total, 65 weights which are constants that define the neural network are shown in the output. That is, if a neural network has ***n*** inputs, ***h*** hidden nodes and ***o*** outputs, it will have (n * h) + h + (h * o) + o weights. In our case, our final neural network has six inputs, five hidden nodes, and five outputs. Therefore, (6 * 5) + 5 + (5 * 5) + 5 will give 65 weights.

We can also visualize the ANN model structure (Fig. 4.9) using ***plotnet()*** function from the **NeuralNetTools** package.

```
> plotnet(annFit$finalModel)
```

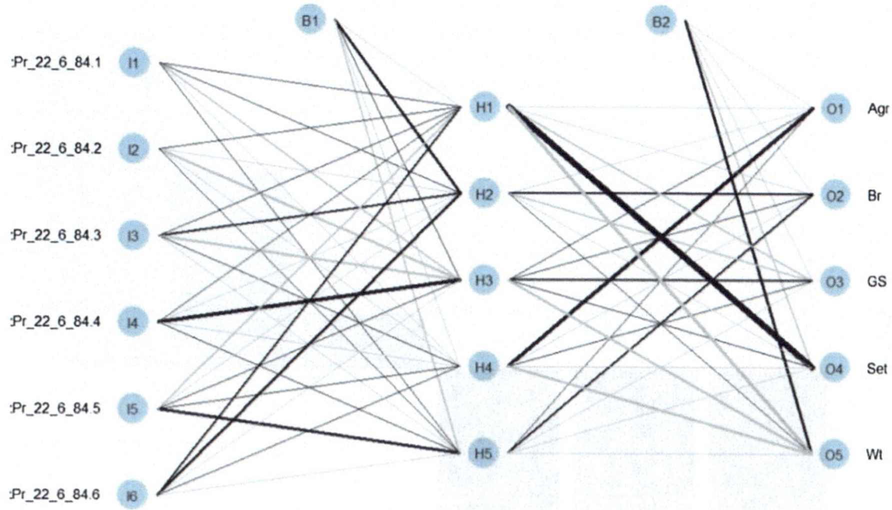

Fig. 4.9 ANN model structure

Next, let's display variable importance using the ***olden()*** function (Fig. 4.10).

```
> olden(annFit)
```

The relative importance of the predictor variables for the ANN model is computed as the sum of the product of raw input-hidden, hidden-output connection weights (Olden et al. 2004). Figure 4.10 shows that band 5, 4 and 7 are the most important predictor variables. For the ANN model, the analysis of the variable importance provides important insights. First, band 5 has the greatest contribution because it is better at separating land use/cover classes in the test site (see Fig. 4.1). Second, we can observe that the negative contribution of bands 1, 2 and 3 is related to high between predictor correlations. Remember, the correlation matrix indicated that bands 1 and 2, and bands 1 and 3 are highly correlated (see Fig. 4.5a–c).

Next, let's perform accuracy assessment. First, we will use the ANN model for prediction and then after build a confusion matrix as shown in the commands below.

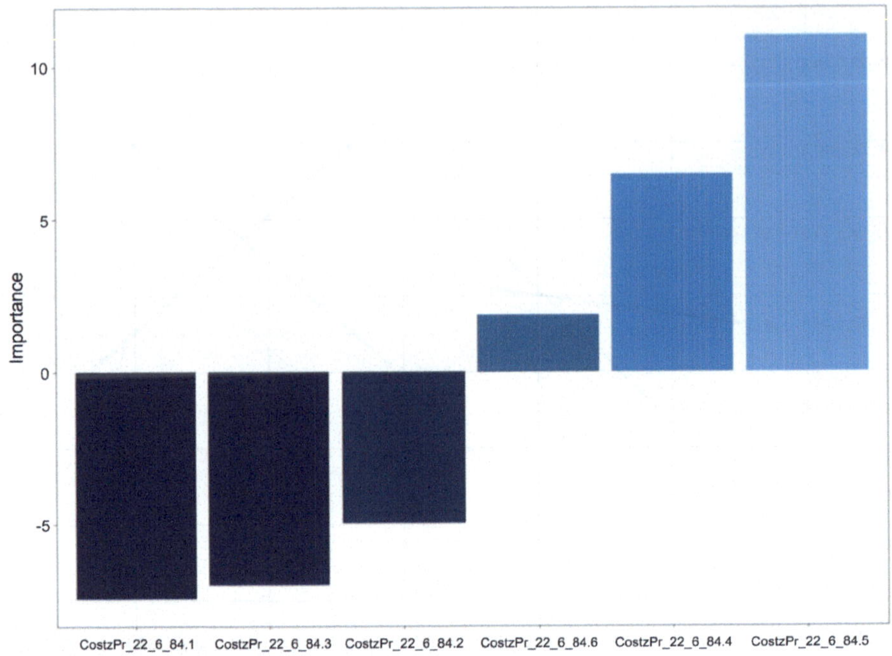

Fig. 4.10 Variable importance for the ANN model

```
> pred_annFit<- predict(annFit, newdata = testing)

> confusionMatrix(data = pred_annFit, testing$Cl1984)

Confusion Matrix and Statistics
          Reference
Prediction Agr  Br  GS Set  Wt
       Agr 249 144  24  12   0
       Br  200 416  39  62   1
       GS   36  15  67  29   3
       Set   7  54   3 174   0
       Wt    0   0   0   0  12


Overall Statistics
               Accuracy : 0.5934
                 95% CI : (0.5684, 0.618)
    No Information Rate : 0.4066
    P-Value [Acc > NIR] : < 2.2e-16
                  Kappa : 0.4083
 Mcnemar's Test P-Value : NA
```

```
Statistics by Class:

                     Class: Agr Class: Br Class: GS Class: Set Class: Wt
Sensitivity              0.5061    0.6614   0.50376     0.6282  0.750000
Specificity              0.8294    0.6710   0.94130     0.9496  1.000000
Pos Pred Value           0.5804    0.5794   0.44667     0.7311  1.000000
Neg Pred Value           0.7826    0.7431   0.95276     0.9213  0.997394
Prevalence               0.3180    0.4066   0.08597     0.1791  0.010343
Detection Rate           0.1610    0.2689   0.04331     0.1125  0.007757
Detection Prevalence     0.2773    0.4641   0.09696     0.1538  0.007757
Balanced Accuracy        0.6677    0.6662   0.72253     0.7889  0.875000
```

In general, the confusion matrix shows high misclassification errors. The overall accuracy is relatively poor. Generally, the KNN classifier has better accuracy than the ANN classifier.

In terms of the producer's accuracy, sensitivity for the agriculture (agr) class is 50%, while the specificity is 82%. This means that the ANN classifier correctly identified agriculture (Agr) 50% of the time, while 82% of the time, the ANN classifier correctly identified non-agriculture (non-agr) classes (that is bareland, green spaces, settlement and water). However, the sensitivity and specificity for the bareland is much closer. It also interesting to note that sensitivity is lower than specificity for the green space class. The positive predicted values (Pos Pred Value) and negative predicted values (Neg Pred Value) shows a similar pattern for all land use/cover classes except water class. Generally, the positive predicted values (Pos Pred Value) for all land use/cover classes except the water class are lower than the negative predicted values (Neg Pred Value). This means that from the map user's perspective, the ANN classifier is poor at identifying the positive land use/cover classes.

The ANN classifier is also better at detecting the true negative classes. A closer look at the confusion matrix shows high misclassification errors, which suggests significant spectral confusion between the agriculture and bareland classes. As a result, the ANN classifier has difficulty to separate agriculture and bareland areas. Here, we see that poor performance of the ANN model is due to high between predictor correlations. Be aware that the ANN model is sensitive to high predictor collinearity.

Step 12: Train single DT classifier

Next, we are going to train a single DT classifier using the CART algorithm available in **rpart** package. The single DT is trained as follows. First a single variable that best splits the training data into two groups is selected. After that, the process is applied separately to each sub-group recursively until the sub-groups reach a minimum size. The second stage of the procedure consists of using cross-validation to prune back the full tree. Here, a "complexity parameter", is defined as the primary tuning parameters. The complexity parameter (cp) is used to control the size of the decision tree and to select the optimal tree size.

For training a single DT classifier, we specify the ***train()*** method as we have done before. We define the response (target) variable as "cl1984 ~ .," the "data" as training, and tune control ("trControl") as "fitControl". Here, we specify the "method" as "rpart" since we are running a single decision tree classifier. All results will be saved in the object "cart_model".

```
> set.seed(hre_seed)
> cart_model<-train(Cl1984~.,data=training,
                    method="rpart",
                    trControl=fitControl)
```

Next, let's check the performance of the classifier. Again we use the ***print()*** and ***plot()*** functions as shown in the commands below.

```
> print(cart_model)

> plot(cart_model)
```

```
CART
6195 samples
   6 predictor
   5 classes: 'Agr', 'Br', 'GS', 'Set', 'Wt'

No pre-processing
Resampling: Cross-Validated (5 fold, repeated 5 times)
Summary of sample sizes: 4958, 4956, 4955, 4956, 4955,
4956, ...
```

```
Resampling results across tuning parameters:
  cp          Accuracy   Kappa
  0.02557127  0.5257145  0.30095325
  0.04651795  0.4817431  0.23776186
  0.05767138  0.4244082  0.06221935

Accuracy was used to select the optimal model using the
 largest value.
The final value used for the model was cp = 0.02557127.
```

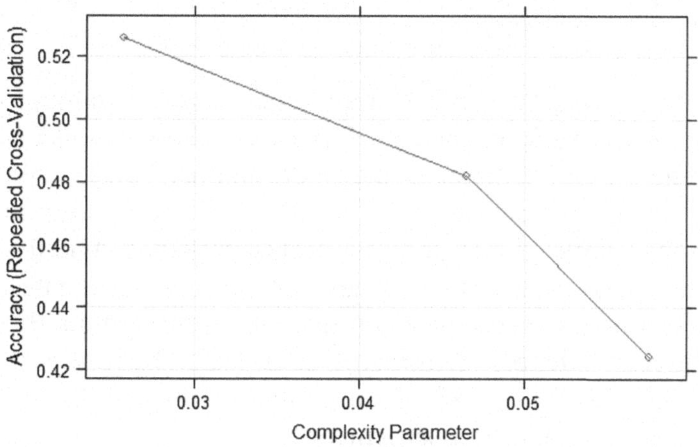

Fig. 4.11 Repeated cross-validation (based on overall accuracy) profile for the DT model

The results show that 6,195 training samples were used for training. We also have six Landsat 5 TM bands and five land use/cover classes. Note that pre-processing was not done since the single DT does not require it. The cross-validation results also give us the sample sizes and the final values that were selected for the best model. The optimal model has cp value of about 0.026 and an accuracy of 52%, which is relatively low (Fig. 4.11).

Moving on, we check the parameters of the best model.

```
> cart_model$finalModel
n= 6195
node), split, n, loss, yval, (yprob)
      * denotes terminal node
```

```
1) root 6195 3676 Br (0.32 0.41 0.086 0.18 0.01)
  2) CostzPr_22_6_84.1< 0.07369175 4010 2420 Agr (0.4 0.
37 0.13 0.085 0.015)
    4) CostzPr_22_6_84.6>=0.1963643 3294 1828 Agr (0.45
 0.41 0.07 0.066 0.0046) *
    5) CostzPr_22_6_84.6< 0.1963643 716  421 GS (0.17 0.
18 0.41 0.17 0.063) *
  3) CostzPr_22_6_84.1>=0.07369175 2185 1160 Br (0.17 0.
47 0.0046 0.35 0.0018)
    6) CostzPr_22_6_84.5>=0.4281517 1730  764 Br (0.21
0.56 0.0058 0.22 0.0017) *
    7) CostzPr_22_6_84.5< 0.4281517 455   68 Set (0.018
 0.13 0 0.85 0.0022) *
```

From the output, "n = ***6195***" represents the total training dataset, while the "***node, split, n, loss, yval, (yprob)***" shows the decision tree legend. The "node" represents the node number, while "split" indicates the feature which was split on. The "n" shows the number of observations, while "loss" shows misclassifications. Finally, "yval" shows the class predictions when processing terminates, while "yprob" shows the class probabilities. As indicated in the results output above "*" denotes terminal node or leaf. The results show that the single DT has four terminal leaves that correspond to agriculture, green spaces, bareland and settlement.

Next, let's visualize the tree using ***rpart.plot()*** function from the **rpart.plot** package (Fig. 4.12).

```
> rpart.plot(cart_model$finalModel)
```

Figure 4.12 shows a very simple classification tree. The tree is a set of binary decisions and terminal nodes connected by branches. The root (first decision) node evaluates the rule as follows. If band 1 (CostzPr_22_6_84.1) is less than the reflectance value of 0.074 and band 7 (CostzPr_22_6_84.6) is greater than or equal to reflectance value 0.2, then a terminal node is reached. The nodes are assigned to the agriculture (agr) and green space (GS) classes. If band 1 (CostzPr_22_6_84.1) is greater than the reflectance value of 0.074, then another binary decision node is evaluated and the tree continues to grow until all branches end with terminal nodes. Note that the DT classifier has completely left out the water class, which indicates classification problems. This is attributed to a number of problems.

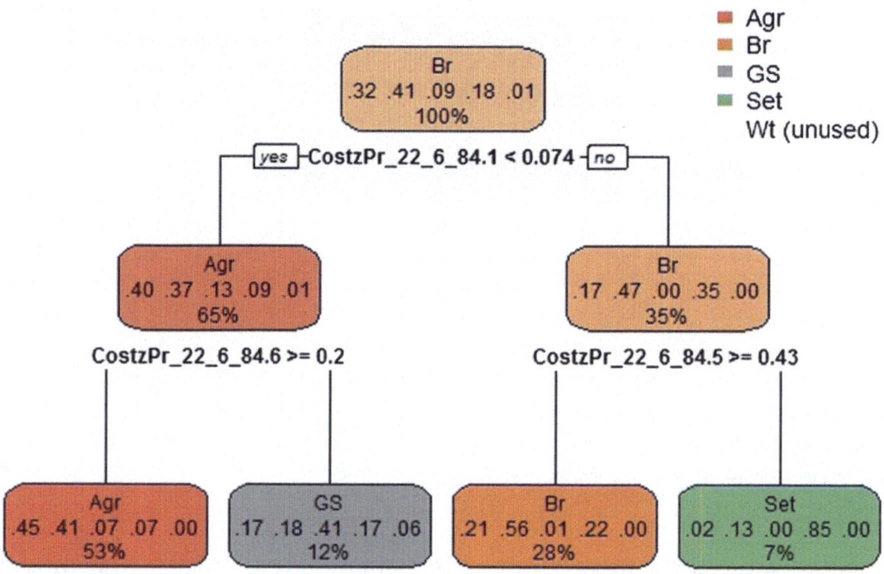

Fig. 4.12 The structure of the optimal singe DT model

First, the water class has a significantly small size of only 64 samples from a training sample size of 6,195 (see summary statistics in step 7). However, single DT requires large training samples for accurate tree classification. Second, the stability of trees is affected by outliers. As observed before (Fig. 4.3), the training dataset has many outliers, which greatly affects the performance of the singe DT classifier.

Next, let's display variable importance using ***varImp()*** function (Fig. 4.13).

```
> cart_varImp <- varImp(cart_model, compete = FALSE)
> ggplot(cart_varImp)
```

Figure 4.13 shows that Landsat 5 TM bands 5, 1 and 7 are the most informative predictor variables. As shown in Fig. 4.1, band 5 is better at separating land use/cover classes in the test site. However, the importance of band 1 is questionable here. This implies instability of the single DT model.

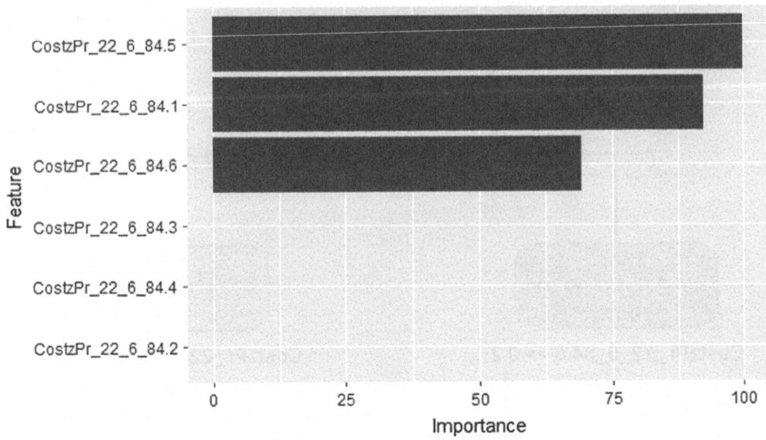

Fig. 4.13 Variable importance for the DT model

Next, let's perform accuracy assessment. First, we use the model for prediction, and then after build a confusion matrix as shown in the commands below. Note the ***predict()*** function will not work here if there are missing values. Therefore use "na.action = na.pass".

```
> pred_cart<- predict(cart_model,newdata = testing,
                      na.action = na.pass)
> confusionMatrix(data = pred_cart, testing$Cl1984)

Confusion Matrix and Statistics
          Reference
Prediction Agr  Br  GS Set  Wt
       Agr 351 363  58  62   0
       Br   98 213   6  99   0
       GS   40  33  69  36  16
       Set   3  20   0  80   0
       Wt    0   0   0   0   0


Overall Statistics
               Accuracy : 0.4609
                 95% CI : (0.4358, 0.4861)
    No Information Rate : 0.4066
    P-Value [Acc > NIR] : 8.619e-06
                  Kappa : 0.226
 Mcnemar's Test P-Value : NA
```

```
Statistics by Class:
                     Class: Agr Class: Br Class: GS Class: Set Class: Wt
Sensitivity              0.7134    0.3386   0.51880    0.28881   0.00000
Specificity              0.5422    0.7789   0.91160    0.98189   1.00000
Pos Pred Value           0.4209    0.5120   0.35567    0.77670       NaN
Neg Pred Value           0.8022    0.6322   0.95270    0.86357   0.98966
Prevalence               0.3180    0.4066   0.08597    0.17906   0.01034
Detection Rate           0.2269    0.1377   0.04460    0.05171   0.00000
Detection Prevalence     0.5391    0.2689   0.12540    0.06658   0.00000
Balanced Accuracy        0.6278    0.5587   0.71520    0.63535   0.50000
```

As expected, the confusion matrix shows no results for the water class given the small training size of the water class. In addition, there are relatively high misclassification errors, especially for the agriculture, bareland and settlement classes. As a result, the overall accuracy is very low given the poor model training results. The "No Information Rate" is about 41% compared to an overall accuracy rate of 46%, which shows poor results.

The class accuracy metrics for the DT classifier shows interesting patterns. In terms of the producer's accuracy, sensitivity is greater than specificity for the agriculture (agr) class. This means that the DT classifier correctly identified agriculture (Agr) 71% of the time, while 54% of the time, the DT classifier correctly identified non-agriculture (non-agr) classes (that is bareland, green spaces, settlement and water). In contrast, sensitivity is less than specificity for the other remaining land use/cover classes (bareland, green spaces, settlement and water). This indicates that the DT classifier is better at detecting negative classes (non-bareland) than the positive classes (e.g., barelend). However, the positive predicted values (Pos Pred Value) and negative predicted values (Neg Pred Value) shows a similar pattern for all land use/cover classes except water class. Generally, the positive predicted values (Pos Pred Value) for all land use/cover classes except the water class are lower than the negative predicted values (Neg Pred Value). This means that from the map user's perspective, the DT classifier is poor at identifying the positive land use/cover classes. In other words, the DT classifier is better at detecting the true negative rate.

The confusion matrix reveals substantial false negatives. There are 351 true agriculture pixels versus 483 non-agriculture pixels. In particular, there is significant confusion between the agriculture and bareland classes. Generally, the single DT accuracy is very poor. This is not surprising given the significant class imbalance. For example, green space and water classes have low proportions compared to agriculture, bareland and settlement classes. In addition, the single DT shows significant instability given the presence of many outliers in the training data (Fig. 4.3). Therefore, the single DT is not effective for image classification in this case.

Step 13: Train radial basis SVM classifier

Many studies have demonstrated the effectiveness of SVMs for remote sensing image classification (Pal and Mather 2005; Nemmour and Chibani 2006). However, SVMs as other machine learning classifiers require proper tuning in order to reduce or eliminate problems such as overfitting. This requires the selection of an appropriate kernel function and cost parameter. In this workbook, we are going to use the radial basis function, which has been successfully used for remote sensing image classification. Automatic tuning will be done to find the optimal cost parameter. Note that all predictors (that is, the Landsat bands) should be centered and scaled before in order to reduce the impact of predictor variables with high attribute values (Kuhn and Johnson 2016).

First, let's train the radial basis SVM classifier.

```
> set.seed(hre_seed)
> svm_model<-train(Cl1984~.,data=training,
                   method = "svmRadial",
                   trControl = fitControl,
                   preProc = c("center", "scale"),
                   tuneLength = 3)

Attaching package: 'kernlab'
The following object is masked from 'package:ggplot2':
    alpha
The following objects are masked from 'package:raster':
    buffer, rotated
```

Next, let's check the performance of the classifier. Again we use the ***print()*** and ***plot()*** functions as shown in the commands below.

```
> print(svm_model)
> plot(svm_model)

Support Vector Machines with Radial Basis Function Ker-
nel
6195 samples
   6 predictor
   5 classes: 'Agr', 'Br', 'GS', 'Set', 'Wt'
Pre-processing: centered (6), scaled (6)
Resampling: Cross-Validated (5 fold, repeated 5 times)
Summary of sample sizes: 4958, 4956, 4955, 4956, 4955,
4956, ...
Resampling results across tuning parameters:
  C     Accuracy   Kappa
  0.25  0.6458140  0.4806940
  0.50  0.6505917  0.4888782
  1.00  0.6555623  0.4963734

Tuning parameter 'sigma' was held constant at a value
of 0.6136129
Accuracy was used to select the optimal model using the
largest value.
The final values used for the model were sigma =
0.6136129 and C = 1.
```

The results show that 6,195 training samples were used for training. The predictors are the six Landsat 5 TM bands, while the five land use/cover classes represent the response (target) variable. Figure 4.14 shows that the optimal model has a Cost value of about 1 and an accuracy of 65%, which is relatively good.

Next, check the parameters of the best model.

```
> svm_model$finalModel

Support Vector Machine object of class "ksvm"

SV type: C-svc  (classification)
 parameter : cost C = 1

Gaussian Radial Basis kernel function.
 Hyperparameter : sigma =  0.613612903068469
Number of Support Vectors : 4632

Objective Function Value : -2839.213 -665.292 -631.9221
 -59.2468 -593.0475 -1008.353 -62.6581 -330.1879 -52.47
98 -51.4142
Training error: 0.317998
```

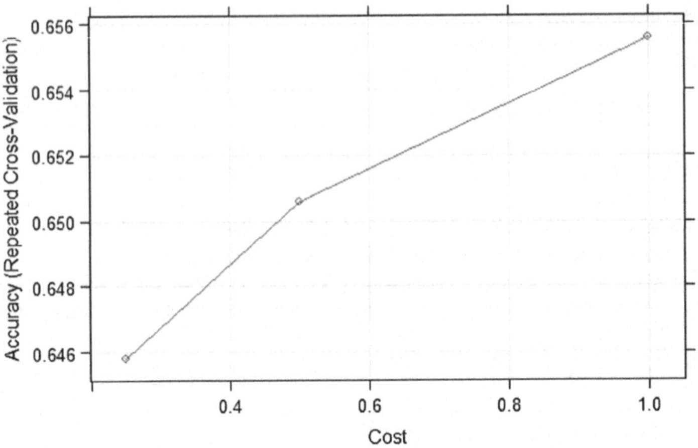

Fig. 4.14 Repeated cross-validation (based on overall accuracy) profile for the SVM model

The output shows that only 4,626 support vectors were selected from a total of 6,195 training samples.

Next, let's display variable importance using ***varImp()*** function (Fig. 4.15).

```
> svm_varImp <- varImp(svm_model, compete = FALSE)
> plot(svm_varImp)
```

Figure 4.15 shows that all Landsat 5 TM bands expect band 4 contribute significantly to the settlement class. This is the same pattern we observed with the KNN model variable importance. As we observed before, band 4 exhibits significant spectral confusion between settlement class, and the agriculture and bareland classes (see Fig. 4.1). Nonetheless, Fig. 4.1 also shows that band 7 (CostzPr_22_6_84.6) exhibit significant spectral confusion between settlement class, and the agriculture and bareland classes. Therefore, it is difficult to understand how the radial basis SVM classifier judged that band 7 has more contribution than band 4. Again, the analysis results are a bit misleading taking into consideration the geographical characteristics of the test site.

Next, let's perform accuracy assessment as we have done in the previous steps. First, we use the SVM model results for prediction, and then build a confusion matrix as shown in the commands below.

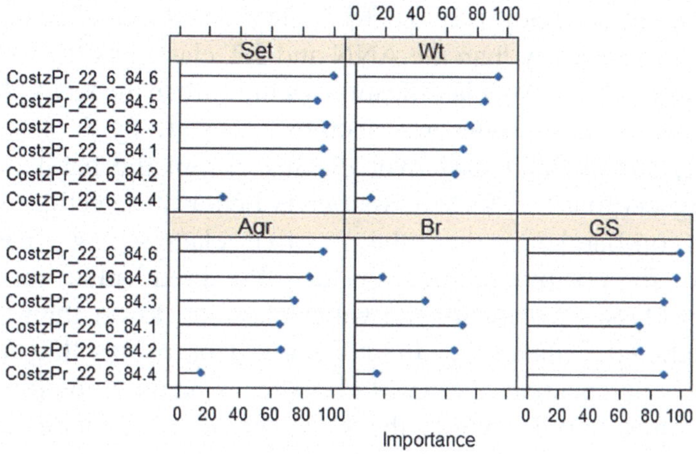

Fig. 4.15 Variable importance for the SVM model

```
> pred_svm<- predict(svm_model, newdata = testing)

> confusionMatrix(data = pred_svm, testing$cl1984)

Confusion Matrix and Statistics
          Reference
Prediction Agr  Br  GS Set  Wt
       Agr 277 123  17  17   0
       Br  185 450  41  54   2
       GS   26  16  71  18   1
       Set   4  40   4 188   1
       Wt    0   0   0   0  12

   Overall Statistics
                        Accuracy : 0.6451
                          95% CI : (0.6207, 0.669)
             No Information Rate : 0.4066
             P-Value [Acc > NIR] : < 2.2e-16
                           Kappa : 0.4808
          Mcnemar's Test P-Value : NA

   Statistics by Class:
                         Class: Agr Class: Br Class: GS Class: Set Class: Wt
   Sensitivity               0.5630    0.7154   0.53383     0.6787  0.750000
   Specificity               0.8512    0.6928   0.95686     0.9614  1.000000
   Pos Pred Value            0.6382    0.6148   0.53788     0.7932  1.000000
   Neg Pred Value            0.8068    0.7804   0.95618     0.9321  0.997394
   Prevalence                0.3180    0.4066   0.08597     0.1791  0.010343
   Detection Rate            0.1791    0.2909   0.04590     0.1215  0.007757
   Detection Prevalence      0.2805    0.4732   0.08533     0.1532  0.007757
   Balanced Accuracy         0.7071    0.7041   0.74535     0.8201  0.875000
```

The overall accuracy is about 65%. The SVM classifier has better classification accuracy than the ANN and DT classifiers. However, the SVM classifier has lower classification accuracy than the KNN classifier.

In terms of the producer's accuracy, sensitivity is lower than specificity for all land use/cover classes except the bareland class. This suggests that the SVM classifier is better at detecting negative classes (non-bareland) than the positive classes (e.g., barelend). Generally, the positive predicted values (Pos Pred Value) for all land use/cover classes except for the water class are lower than the negative predicted values (Neg Pred Value). This means that from the map user's perspective, the SVM classifier is better at identifying the negative rates. In other words, the SVM classifier is better at detecting non-agriculture or non-bareland classes.

The confusion matrix shows significant confusion between the agriculture and bareland classes. This indicates that the radial basis SVM classifier has difficulty separating the agriculture and bareland classes. However, the radial basis SVM classifier is much better at handling outliers compared to the single DT and the ANN classifiers.

Step 14: Train the random forest (RF) classifier

Random forests have also been used successfully for remote sensing image classification (Lu and Weng 2007; Mellor et al. 2013). For RF models, the most important tuning parameter is mtry, which represents the number of randomly selected predictors (k) to choose from each split.

Let's start to train the RF model.

```
> set.seed(hre_seed)
> rf_model<-train(Cl1984~.,data=training, method="rf",
                  trControl=fitControl,
                   prox=TRUE,
                   fitBest = FALSE,
                   returnData = TRUE)
randomForest 4.6-12
Type rfNews() to see new features/changes/bug fixes.
Attaching package: 'randomForest'
The   following   object   is   masked   from   'pack-
age:gridExtra':
    combine
The following object is masked from 'package:ggplot2':
    Margin
```

Next, let's check the RF model performance. Again we use the ***print()*** and ***plot()*** functions as shown in the commands below.

```
> print(rf_model)
> plot(rf_model)
```

```
Random Forest
6195 samples
   6 predictor
   5 classes: 'Agr', 'Br', 'GS', 'Set', 'Wt'
No pre-processing
Resampling: Cross-Validated (5 fold, repeated 5 times)
Summary of sample sizes: 4958, 4956, 4955, 4956, 4955,
4956, ...
Resampling results across tuning parameters:

  mtry  Accuracy   Kappa
  2     0.7412747  0.6248495
  4     0.7368197  0.6188088
  6     0.7339776  0.6146914
Accuracy was used to select the optimal model using the
 largest value.
The final value used for the model was mtry = 2.
```

The results show that 6,195 training samples were used for training. The six predictor variables are the Landsat 5 TM bands, while the five land use/cover classes represent the response (target) variable. Note that pre-processing was not done since the RF classifier requires minimum pre-processing. The RF model was tuned over the values of the mtry parameter. The best model had an mtry value of 2 with an overall accuracy of 74%, which is relatively good (Fig. 4.16).

Next, let's check the parameters of the best model.

```
> rf_model$finalModel
```

```
Call:
 randomForest(x = x, y = y, mtry = param$mtry, proximit
y = TRUE,       fitBest = FALSE, returnData = TRUE)
               Type of random forest: classification
                     Number of trees: 500
No. of variables tried at each split: 2
        OOB estimate of  error rate: 24.6%
```

```
Confusion matrix:
     Agr   Br  GS Set Wt class.error
Agr 1458  440  37  34  0   0.2595226
Br   434 1899  40 145  1   0.2461294
GS    42   67 413  13  0   0.2280374
Set   44  160  39 865  0   0.2193141
Wt     8   16   1   3 36   0.4375000
```

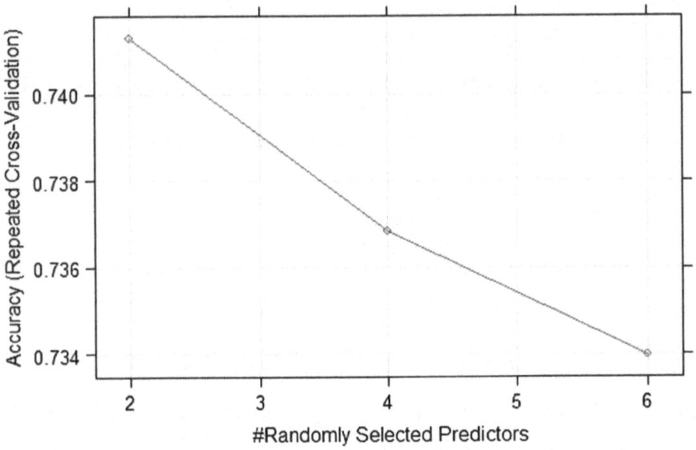

Fig. 4.16 Repeated cross-validation (based on overall accuracy) profile for the RF model

The output shows a confusion matrix for the best model (after cross-validation). A total of 500 decision trees were used in the RF model. From the six predictor variables (six Landsat 5 TM bands), only two predictor variables (bands) were selected at each split. The out-of-bag (OOB) estimate of error rate is 24.6%.

Next, let's display variable importance using ***varImp()*** function.

```
> rf_varImp <- varImp(rf_model, compete = FALSE)
> plot(rf_varImp)
```

Figure 4.17 shows the relative importance of the contribution of the six Landsat 5 TM bands. Band 5 has great contributions followed by bands 4 and 7. The results provide valuable insight of the RF model performance. For example, Fig. 4.1 shows that band 5 was better at separating land use/cover classes. In this case, the RF correctly managed to identify band 5 as the most informative predictor variable.

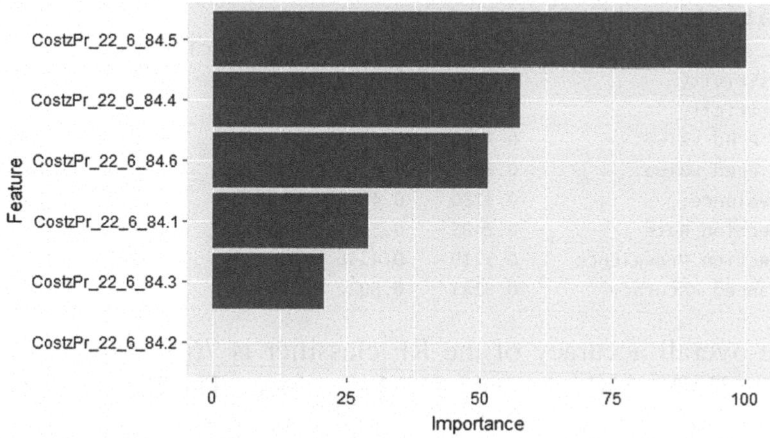

Fig. 4.17 Variable importance for the RF model

Next, let's perform accuracy assessment as we have done in the previous steps. First, we use the RF model results for prediction, and then build a confusion matrix as shown in the commands below.

```
> pred_rf <- predict(rf_model$finalModel,
            newdata = testing)

> confusionMatrix(data = pred_rf, testing$Cl1984)

Confusion Matrix and Statistics
          Reference
Prediction Agr  Br  GS Set  Wt
       Agr 372  96  18  11   1
       Br  103 487  13  39   1
       GS   11   7  94  14   0
       Set   6  37   4 213   2
       Wt    0   2   4   0  12


Overall Statistics
                Accuracy : 0.7615
                  95% CI : (0.7394, 0.7825)
     No Information Rate : 0.4066
     P-Value [Acc > NIR] : < 2e-16
                   Kappa : 0.6549
  Mcnemar's Test P-Value : 0.05252
```

```
Statistics by Class:

                     Class: Agr Class: Br Class: GS Class: Set Class: Wt
Sensitivity              0.7561    0.7742   0.70677     0.7690  0.750000
Specificity              0.8806    0.8301   0.97737     0.9614  0.996081
Pos Pred Value           0.7470    0.7574   0.74603     0.8130  0.666667
Neg Pred Value           0.8856    0.8429   0.97255     0.9502  0.997384
Prevalence               0.3180    0.4066   0.08597     0.1791  0.010343
Detection Rate           0.2405    0.3148   0.06076     0.1377  0.007757
Detection Prevalence     0.3219    0.4156   0.08145     0.1694  0.011635
Balanced Accuracy        0.8183    0.8022   0.84207     0.8652  0.873040
```

The overall accuracy of the RF classifier is 76%, which is better than the KNN, ANN, DT and SVM classifiers. The RF classifier also performs better in terms of individual class performance. The producer's accuracy (sensitivity and specificity) and the user's accuracy (positive predicted values and negative predicted values) are relatively better for all land use/cover classes. In addition, the RF classifier is better at handling outliers than the other machine learning classifiers used in this workbook. While accuracy assessment results are relatively good, misclassification problems are still apparent. For example, 103 bareland pixels were misclassified as agriculture. This means there is still need to improve land use/cover classification accuracy further.

Step 15: Compare all the machine learning classifiers

Let's proceed to compare all machine learning classifiers based on the cross-validation statistics. Here, we are going to use the ***resamples()*** function because the machine learning classifiers share a common set of resampled data sets. First, we run the ***resamples()***, and then check the resamps object as shown in the command below.

```
> resamps <- resamples(list(kknn = knnFit,
                            nnet = annFit,
                            rpart = cart_model,
                            kernlab = svm_model,
                            e1071 = rf_model))

> resamps
```

```
Call:
resamples.default(x = list(kknn = knnFit, nnet = annFit,
 rpart = cart_model, kernlab = svm_model, randomForest
= rf_model))
Models: kknn, nnet, rpart, kernlab, randomForest
Number of resamples: 25
Performance metrics: Accuracy, Kappa
Time estimates for: everything, final model fit
```

Next, let's check the summary statistics, and then display the results in graphic form. Note that this is a comparison of the model training performance.

```
> summary (resamps)
> bwplot(resamps, layout = c(3, 1))

Call:
summary.resamples(object = resamps)
Models: kknn, nnet, rpart, kernlab, randomForest
Number of resamples: 25
Accuracy
```

	Min.	1st Qu.	Median	Mean	3rd Qu.	Max.	NA's
kknn	0.6838710	0.6927419	0.7046005	0.7033111	0.7102502	0.7231638	0
nnet	0.5940274	0.6101695	0.6161290	0.6156594	0.6222760	0.6384181	0
rpart	0.4786118	0.5161551	0.5327405	0.5257145	0.5375303	0.5512510	0
kernlab	0.6319613	0.6483871	0.6537530	0.6555623	0.6594027	0.6844229	0
randomForest	0.7197092	0.7360775	0.7419355	0.7412747	0.7481840	0.7619048	0

The summary statistics and Fig. 4.18 show that the performance of the machine learning classifiers varies greatly. Clearly, the RF classifier has the highest performance, while the single DT tree (rpart) classifier has the lowest performance. However, it should be noted that it took more time to tune the RF classifier. Interestingly, the simple KNN classifier performed better than the advanced radial basis SVM classifier. This shows that it is better to start training simple machine learning classifiers, which are less computationally intensive.

Step 16: Perform classification

Next, let's perform land use/cover classification using all machine learning classifiers. Be patience, since it takes time to run all the classifications.

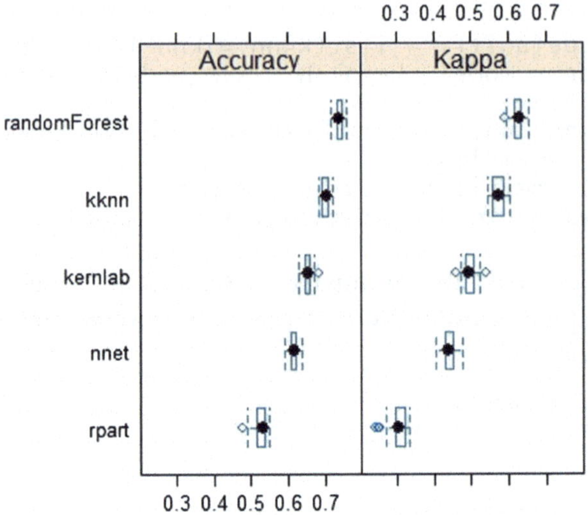

Fig. 4.18 Model validation results from different machine learning classifiers

```
> timeStart<- proc.time() # measure computation time
> LC_knnFit_84 <-predict(rasVars,knnFit)
> LC_ann_84 <-predict(rasVars,annFit)
> LC_cart_84 <-predict(rasVars,cart_model)
> LC_svm_84 <-predict(rasVars,svm_model)
> LC_rf_84 <-predict(rasVars,rf_model)
> proc.time() - timeStart
user  system elapsed
1182.86       51.46     1242.01
```

Step 17: Display final land use/cover classifications

We are going to use the ***ggplot2()*** function to display all land use/cover maps using the following commands.

```
> LC_knnFit_84a <- gplot(LC_knnFit_84) + geom_raster(ae
s(fill = factor(value, labels=c("Agriculture", "Barelan
d", "Green Spaces", "Urban", "Water")))) + scale_fill_m
anual(values = c("yellow", "grey", "green3", "red", "bl
ue3"), name= "Land Cover") + ggtitle("K Nearest Neighbo
ur Classification") +theme(plot.title = element_text(li
neheight=.4, face="bold")) + coord_equal()
```

```
> LC_ann_84a <- gplot(LC_ann_84) + geom_raster(aes(fill
 = factor(value, labels=c("Agriculture", "Bareland", "
Green Spaces", "Urban", "Water")))) + scale_fill_manual
(values = c("yellow", "grey", "green3", "red", "blue3"),
 name= "Land Cover") + ggtitle("Artificial Neural Netwo
rk Classification") + theme(plot.title = element_text(l
ineheight=.4, face="bold")) + coord_equal()

> LC_cart_84a <- gplot(LC_cart_84) + geom_raster(aes(fi
ll = factor(value, labels=c("Agriculture", "Bareland",
"Green Spaces", "Urban")))) + scale_fill_manual(values
= c("yellow", "grey", "green3", "red"), name= "Land Cov
er") + ggtitle("Decision Trees Classification") +theme
(plot.title = element_text(lineheight=.4, face="bold"))
 + coord_equal()

> LC_svm_84a <- gplot(LC_svm_84) + geom_raster(aes(fill
 = factor(value, labels=c("Agriculture", "Bareland",
"Green Spaces", "Urban", "Water")))) +scale_fill_manual
(values = c("yellow", "grey", "green3", "red", "blue3"),
 name= "Land Cover") + ggtitle("Support Vector Machine
 Classification") +theme(plot.title = element_text(line
height=.4, face="bold")) + coord_equal()

> LC_rf_84a <- gplot(LC_rf_84) + geom_raster(aes(fill =
 factor(value, labels=c("Agriculture", "Bareland", "Gre
en Spaces", "Urban", "Water")))) + scale_fill_manual(va
lues = c("yellow", "grey", "green3", "red", "blue3"),
name= "Land Cover") + ggtitle("Random Forest Classifica
tion") +theme(plot.title = element_text(lineheight=.4,
face="bold")) + coord_equal()
```

Finally, let's us arrange the land use/cover results (Fig. 4.19) using ***grid.arrange()*** function from the gridExtra package.

```
> grid.arrange(LC_knnFit_84a, LC_ann_84a, LC_cart_84a,
LC_svm_84a, LC_rf_84a, ncol=2)
```

4.2.4 Summary for Tutorial 1

The RF classifier outperformed the KNN, ANN, DT and radial basis SVM classifiers. In addition, the RF classifier was not strongly affected by outliers and class imbalances problems. Therefore, the RF classifier will be used for image classification in tutorial 2, and Chap. 5.

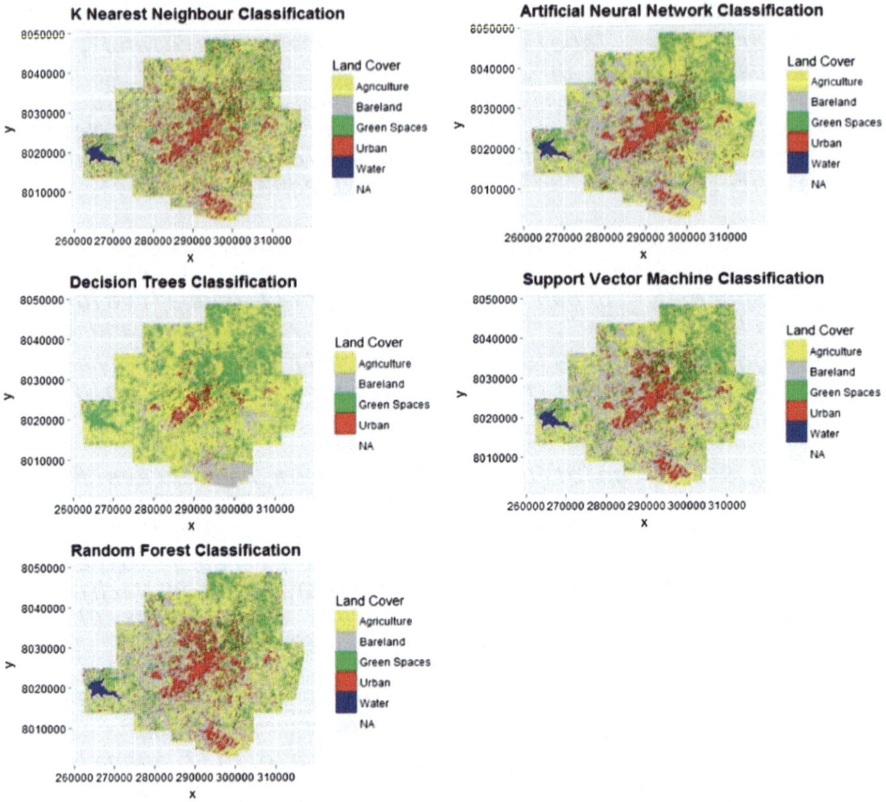

Fig. 4.19 Land use/cover classification

4.3 Tutorial 2: Multidate Landsat Image Classification

In this tutorial, we are going to classify multidate Landsat data set, which comprises Landsat 5 TM imagery acquired during the rainy and dry seasons (Table 1.1 and Fig. 1.1 in Chap. 1). Multidate Landsat data sets have been reported to improve classification accuracy since they account for seasonality or vegetation phenology in the classification.

4.3.1 Objective

- Classify multidate Landsat 5 TM imagery (acquired on 22 June 1984 and 31 December 1984) using the RF classifier.

4.3.2 Procedure

In this tutorial, we are going to use multidate Landsat 5 TM imagery for image classification based on RF classifier. Following are the steps for image classification.

Step 1: Load the necessary packages or libraries

First, we need to load the following libraries.

```
> library(raster)
> library(ggplot2)
> library(reshape)
> library(grid)
> library(gridExtra)
> library(RStoolbox)
> library(caret)
> library(rasterVis)
> library(corrplot)
> library(doParallel)
```

Step 2: Set the working directory and load raster data

Next, set up the working directory where all your datasets are located.

```
> setwd("E:\\DataFolderPath\\~")
```

Create a list of raster bands that will be used for image classification.

```
> rasList <- list.files(getwd(),pattern="img$",
              full.names=TRUE)
```

Combine or stack the raster layers.

```
> rasVars<- stack(rasList)
```

Once we have loaded and created a raster stack, the next step is to check the attributes of the multidate Landsat 5 TM imagery.

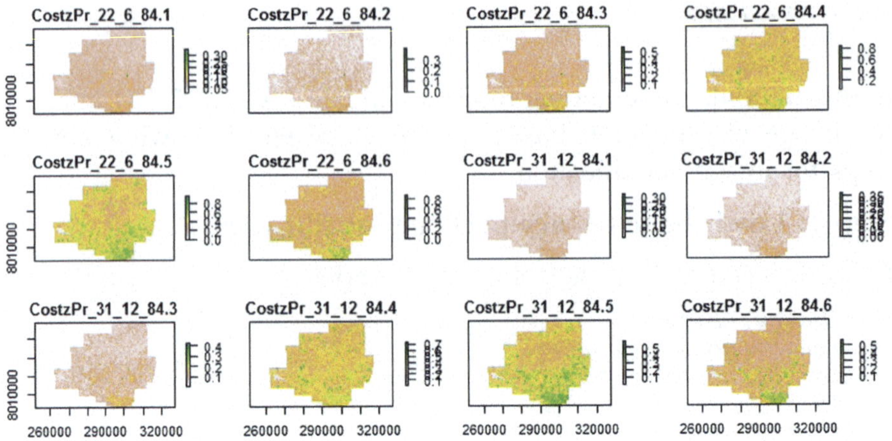

Fig. 4.20 A subset of multidate Landsat 5 TM bands for the test site

```
> rasVars
class       : RasterStack
dimensions  : 1533, 1832, 2808456, 12  (nrow, ncol, ncell, nlayers)
resolution  : 30, 30  (x, y)
extent      : 262065, 317025, 8002765, 8048755  (xmin, xmax, ymin, ymax)
coord. ref. : +proj=utm +zone=36 +south +ellps=WGS84 +towgs84=0,0,0,-0,-0,-0,0 +units=m +no_defs
names: CostzPr_22_6_84.1, CostzPr_22_6_84.2, CostzPr_22_6_84.3, CostzPr_22_6_84.4,CostzPr_22_6_84.5, CostzPr_22_6_84.6,
CostzPr_31_12_84.1, CostzPr_31_12_84.2, CostzPr_31_12_84.3, CostzPr_31_12_84.4,CostzPr_31_12_84.5,CostzPr_31_12_84.6
min values  :   1.090038e-02,  0.000000e+00,  1.815953e-03,  1.600820e-02,  0.000000e+00,  0.000000e+00,
5.680612e-03, 0.000000e+00,   1.842262e-03,   1.985818e-03,   4.932763e-03,   8.515586e-14
max values  :      0.6557847,      0.7136533,      0.8484794,      1.0000000,      1.0000000,      1.0000000,
0.3247315,      0.5033199,      0.6595355,      0.9100135,      0.6457630,      0.9494425
```

Plot the multidate Landsat 5 TM imagery as single bands using ***plot ()*** function (Fig. 4.20).

```
> TM5_Multidate_84 <- plot(rasVars)
```

Step 3: Load the training data

Next, load the training datasets using ***readOGR()*** function.

```
> ta_data <- readOGR(getwd(), "TA_1984")
OGR data source with driver: ESRI Shapefile
Source: "E:/Practical Machine Learning/ML_Woorkbook_Dat
asets/Data/ImageClassification/DS1_LandsatOnly/Dataset2
_Multidate2", layer: "TA_1984"
with 7769 features
It has 1 fields
```

Check the structure of the training data.

```
> str(ta_data)
Formal class 'SpatialPointsDataFrame' [package "sp"] with 5 slots
  ..@ data       :'data.frame':       7769 obs. of  1 variable:
  .. ..$ Cl1984: Factor w/ 5 levels "Agr","Br","GS",..: 4 4 4 4 4 4 4 4 4 4 ...
```

We can also check the number of training points in the point shapefile using the ***summary()*** function.

```
> summary(ta_data)
Object of class SpatialPointsDataFrame
Coordinates:
                min       max
coords.x1  262944.4  315205.6
coords.x2 8002868.9 8047944.6
Is projected: TRUE
proj4string :
[+proj=utm +zone=36 +south +datum=WGS84 +units=m +no_defs +el
lps=WGS84 +towgs84=0,0,0]
Number of points: 7769

Data attributes:
 Cl1984
 Agr:2467
 Br :3169
 GS : 668
 Set:1385
 Wt :  80
```

Step 4: Create a data frame with labeled training points containing all reflectance values

Let's create a new data frame that contains labeled training points with all reflectance values for each land use/cover class. Use the ***data.frame()*** function to create the "ta_data@data" object.

```
> ta<-as.data.frame(extract(rasVars, ta_data))

>ta_data@data=data.frame(ta_data@data,ta[match(rownames
(ta_data@data), rownames(ta)),])
```

Let's examine the structure of the whole training dataset using ***str()*** function. This gives us an overview of the dataset.

```
> str(ta_data@data)
```

The data frame consists of 7,769 training points. The five land use/cover classes ("Cl1984") represent the response variable, while the 12 Landsat 5 TM bands (from the dry and wet seasons) represent the predictor variables.

```
'data.frame':   7769 obs. of  13 variables:
 $ Cl1984              : Factor w/ 5 levels "Agr","Br","G
S",..: 4 4 4 4 4 4 4 4 4 4 ...
 $ CostzPr_22_6_84.1 : num  0.1195 0.1093 0.1059 0.0856 0.1161 ...
 $ CostzPr_22_6_84.2 : num  0.1117 0.083 0.0687 0.0687 0.0973 ...
 $ CostzPr_22_6_84.3 : num  0.182 0.155 0.135 0.128 0.182 ...
 $ CostzPr_22_6_84.4 : num  0.284 0.234 0.209 0.192 0.284 ...
 $ CostzPr_22_6_84.5 : num  0.365 0.276 0.16 0.165 0.276 ...
 $ CostzPr_22_6_84.6 : num  0.338 0.265 0.127 0.16 0.233 ...
 $ CostzPr_31_12_84.1: num  0.1004 0.0878 0.0546 0.1068 0.1068 ...
 $ CostzPr_31_12_84.2: num  0.1031 0.0864 0.0331 0.1065 0.1065 ...
 $ CostzPr_31_12_84.3: num  0.1539 0.129 0.0608 0.1539 0.1601 ...
 $ CostzPr_31_12_84.4: num  0.22 0.189 0.103 0.205 0.232 ...
 $ CostzPr_31_12_84.5: num  0.275 0.216 0.141 0.208 0.242 ...
 $ CostzPr_31_12_84.6: num  0.271 0.218 0.139 0.206 0.214 ...
```

Next, check the summary statistics to see if there are NAs.

```
> summary(ta_data@data)
Cl1984       CostzPr_22_6_84.1 CostzPr_22_6_84.2 CostzPr_22_6_84.3  CostzPr_22_6_84.4
CostzPr_22_6_84.5 CostzPr_22_6_84.6
 Agr:2467   Min.    :0.01769   Min.    :0.00000   Min.    :0.008483   Min.    :0.03276
  Min.    :0.00507   Min.    :0.0000
 Br :3169   1st Qu.:0.05163   1st Qu.:0.02566   1st Qu.:0.101816   1st Qu.:0.29237
 1st Qu.:0.35349   1st Qu.:0.2328
 GS : 668   Median :0.06181   Median :0.04716   Median :0.135149   Median :0.33424   Median
 :0.41986   Median :0.2896
 Set:1385   Mean    :0.06969   Mean    :0.05681   Mean    :0.139687   Mean    :0.33462   Mean
 :0.42028   Mean    :0.2993
 Wt :  80   3rd Qu.:0.08218   3rd Qu.:0.07583   3rd Qu.:0.168482   3rd Qu.:0.37611
 3rd Qu.:0.49175   3rd Qu.:0.3544
```

The statistics show that there are NAs. Therefore, we should remove the NAs before we proceed.

```
> ta_data@data <- na.omit(ta_data@data)

> complete.cases(ta_data@data)

 [1] TRUE TRUE TRUE TRUE TRUE TRUE TRUE TRUE TRUE TRUE TRUE TRUE TRUE TRUE TRUE TRUE TRUE
TRUE TRUE TRUE TRUE TRUE TRUE TRUE TRUE
 [26] TRUE TRUE TRUE TRUE TRUE TRUE TRUE TRUE TRUE TRUE TRUE TRUE TRUE TRUE TRUE TRUE TRUE
TRUE TRUE TRUE TRUE TRUE TRUE TRUE TRUE
 [51] TRUE TRUE TRUE TRUE TRUE TRUE TRUE TRUE TRUE TRUE TRUE TRUE TRUE TRUE TRUE TRUE TRUE
TRUE TRUE TRUE TRUE TRUE TRUE TRUE TRUE
 [76] TRUE TRUE TRUE TRUE TRUE TRUE TRUE TRUE TRUE TRUE TRUE TRUE TRUE TRUE TRUE TRUE TRUE
TRUE TRUE TRUE TRUE TRUE TRUE TRUE TRUE
[101] TRUE TRUE TRUE TRUE TRUE TRUE TRUE TRUE TRUE TRUE TRUE TRUE TRUE TRUE TRUE TRUE TRUE
TRUE TRUE TRUE TRUE TRUE TRUE TRUE TRUE
[126] TRUE TRUE TRUE TRUE TRUE TRUE TRUE TRUE TRUE TRUE TRUE TRUE TRUE TRUE TRUE TRUE TRUE
TRUE TRUE TRUE TRUE TRUE TRUE TRUE TRUE
[151] TRUE TRUE TRUE TRUE TRUE TRUE TRUE TRUE TRUE TRUE TRUE TRUE TRUE TRUE TRUE TRUE TRUE
TRUE TRUE TRUE TRUE TRUE TRUE TRUE TRUE
[176] TRUE TRUE TRUE TRUE TRUE TRUE TRUE TRUE TRUE TRUE TRUE TRUE TRUE TRUE TRUE TRUE TRUE
TRUE TRUE TRUE TRUE TRUE TRUE TRUE TRUE
 [ reached getOption("max.print") -- omitted 6742 entries ]
```

Step 6: Prepare training and test data sets

We will partition the training data set into training and testing sets. First, set a pre-defined value using ***set.seed()***.

```
> hre_seed<- 27
> set.seed(hre_seed
```

Next, split the training dataset into training and test sets using ***createDataPartition()***. Again, 80% of training data set will be used for training, while 20% will be used for testing or accuracy assessment.

```
> inTraining <- createDataPartition(ta_data@data$Cl1984,
                    p = .80, list = FALSE)

> training<- ta_data@data[ inTraining,]

> testing  <- ta_data@data[-inTraining,]
```

Check the dimensions and size of the training and testing sets using the ***dim()*** function.

```
> dim(training)
[1] 6195   13

> dim(testing)
[1] 1547   13
```

The output shows there are 6,195 points for training, 1,547 points for testing, and 13 variables.

Step 7: Check the summary statistics (training and test data sets)

Once again, check the descriptive statistics of the training and testing data set using ***summary()*** function. Here, I will only display a sample of the training set. However, you will see the whole training and testing sets when you execute the ***summary()*** function.

```
> summary(training)
Cl1984      CostzPr_22_6_84.1 CostzPr_22_6_84.2 CostzPr_22_6_84.3  CostzPr_22_6_84.4
 CostzPr_22_6_84.5 CostzPr_22_6_84.6
Agr:1969   Min.    :0.01769   Min.    :0.00000   Min.    :0.008483   Min.    :0.03276
Min.   :0.00507   Min.    :0.0000
Br :2519   1st Qu.:0.05163   1st Qu.:0.02566   1st Qu.:0.101816   1st Qu.:0.29237
 1st Qu.:0.35349   1st Qu.:0.2328
GS : 535   Median :0.06181   Median :0.04716   Median :0.135149   Median :0.33424
Median :0.41986   Median :0.2896
Set:1108   Mean    :0.06989   Mean    :0.05706   Mean    :0.139997   Mean    :0.33493
Mean   :0.42096   Mean    :0.2998
Wt :  64   3rd Qu.:0.08218   3rd Qu.:0.07583   3rd Qu.:0.168482   3rd Qu.:0.37611
3rd Qu.:0.49175   3rd Qu.:0.3544

> summary(testing)
```

Step 8: Visualization

We are going to use ***featurePlot()*** and ***ggplot2()*** functions in order to visualize the training data set. As we have done in previous section, let's display the density plots (Fig. 4.21) using the ***featurePlot()*** function.

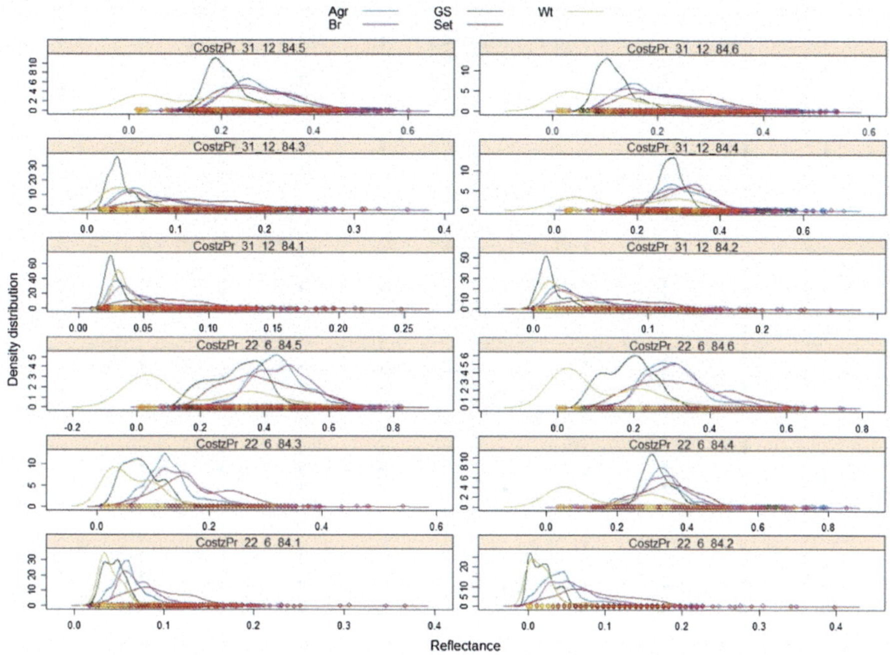

Fig. 4.21 Density plots for the training set

```
> featurePlot(x = training[, 2:13],
          y = training$Cl1984,
          plot = "density",
          labels=c("Reflectance", "Density distribution"),
          scales = list(x = list(relation="free"),
                             y = list(relation="free")),
         layout = c(2, 6),
        auto.key = list(columns = 3))
```

Next, use the ***featurePlot()*** function to display the box plots (Fig. 4.22).

```
> featurePlot(x = training[, 2:7], y = training$Cl1984,
              plot = "box",
              scales = list(y = list(relation = "free"),
                             x = list(rot = 90)),
              layout = c(2,6),
              auto.key = list(colmns = 2))
```

Finally, let's check between predictor correlations for the 12 Landsat 5 TM bands using the ***cor()*** and ***corrplot()*** functions.

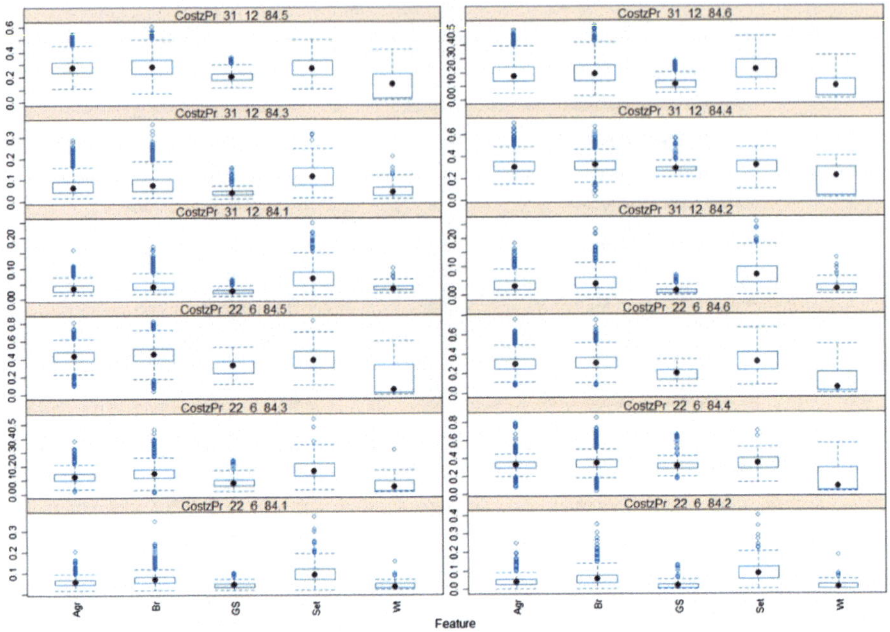

Fig. 4.22 Box plots for the training set

```
> bandCorrelations <- cor(training[, 2:13])

> dim(bandCorrelations)
[1] 12 12
```

This above output shows that we have a correlation matrix of 12 Landsat 5 TM bands (remember, we have combined two Landsat 5 TM imagery from the dry and wet seasons).

```
> bandCorrelations
CostzPr_22_6_84.1 CostzPr_22_6_84.2 CostzPr_22_6_84.3 CostzPr_22_6_84.4 CostzPr_22_6_84.5
CostzPr_22_6_84.6 CostzPr_31_12_84.1
CostzPr_22_6_84.1         1.0000000         0.9610577         0.8966841         0.5365460
        0.5901836         0.7354192         0.7769860
CostzPr_22_6_84.2         0.9610577         1.0000000         0.9580730         0.6419002
        0.6828810         0.8056505         0.7260617
CostzPr_22_6_84.3         0.8966841         0.9580730         1.0000000         0.6560069
        0.7907537         0.8759879         0.6378483
```

```
CostzPr_22_6_84.4          0.5365460          0.6419002          0.6560069          1.0000000
         0.6714087          0.5553501          0.2422702
CostzPr_22_6_84.5          0.5901836          0.6828810          0.7907537          0.6714087
         1.0000000          0.9059390          0.3216360
CostzPr_22_6_84.6          0.7354192          0.8056505          0.8759879          0.5553501
         0.9059390          1.0000000          0.5233310
CostzPr_31_12_84.1         0.7769860          0.7260617          0.6378483          0.2422702
         0.3216360          0.5233310          1.0000000
```

Note that I have reduced the output for illustration purposes only because the correlation coefficients are too long. Therefore, let's create a correlation matrix (Fig. 4.23) using the ***corrplot.mixed()*** function.

```
> corrplot.mixed(bandCorrelations,lower.col="black",
number.cex = .7, upper = "color")
```

Next, let's check the bands that are highly correlated based on a cutoff point of say 0.8.

```
> highCorrelated <- findCorrelation(bandCorrelations,
cutoff = .8)
```

In addition, we can run commands to check how many predictors are highly correlated. More importantly, we can get specific names of the highly correlated predictor variables.

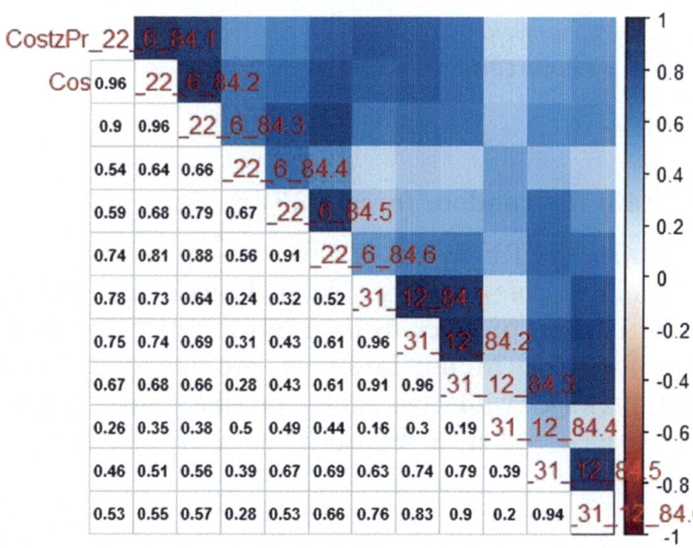

Fig. 4.23 Correlation matrix for the multidate Landsat 5 TM bands

```
> length(highCorrelated)
[1] 6

> names(training)[highCorrelated]
[1] "CostzPr_22_6_84.2"  "CostzPr_22_6_84.1"  "CostzPr_
22_6_84.5"  "CostzPr_31_12_84.1" "CostzPr_31_12_84.2" "
CostzPr_31_12_84.4"
```

Here,we are not removing any highly correlated predictors (bands). This is because the RF model uses a subset of predictors to construct each decision tree. Therefore, correlation between the single decision trees is reduced, thereby improving model performance and accuracy. However, there is an option to remove highly correlated predictors as shown in the command below. This is useful when you are dealing with machine learning methods such as ANN, which are sensitive to high predictor collinearity.

```
> reduced_training <- training[, -highCorrelated]
```

Step 9: Set-up model tuning parameters

As we have done in the previous section, we use the ***trainControl()*** function from **caret** package to set up model tuning parameters in order to measure performance. This will also choose the "optimal" model across these parameters, and estimate model performance from a training set.

```
> fitControl<- trainControl(
   method = "repeatedcv",
   number = 5,
   repeats = 5)
```

Step 10: Train the random forest (RF) model

Before we train the RF model, we are going to use a function to "register" multiple cores (this is a parallel processing technique that specifies the number of cores to be used) in order to increase computational efficiency. In this example, we are going to use the **doParallel** package with five cores on the same machine. The package is loaded and then registered as shown in the commands below.

```
> cl <- makePSOCKcluster(5)
> registerDoParallel(cl)
```

Next, let us train the RF model.

```
> cl <- makePSOCKcluster(5)
> registerDoParallel(cl)


> set.seed(hre_seed)
> rf_model<-train(Cl1984~.,data=training, method="rf",
                  trControl=fitControl,
                  prox=TRUE,
                  fitBest = FALSE,
                  returnData = TRUE)
randomForest 4.6-12
Type rfNews() to see new features/changes/bug fixes.
Attaching package: 'randomForest'
The following object is masked from 'pack-
age:gridExtra':
    combine
The following object is masked from 'package:ggplot2':
    Margin
  > proc.time() - timeStart # user time and system time.
     user        system   elapsed
     30.95       0.42     699.59
```

Note that on an Intel (R) core (TM) processor with a 64-bit Windows operating system and 16 GB memory, it took approximately 12 min to train the RF model. Without parallel processing, it took me approximately 25 min.

After finishing training, we can use the ***stopCluster()*** function to stop the parallel processing.

```
> stopCluster(cl)
```

Next, let's run the ***print()*** and ***plot()*** functions in order to check the model performance.

```
> print(rf_model)
> plot (rf_model)

Random Forest

6195 samples
  12 predictor
   5 classes: 'Agr', 'Br', 'GS', 'Set', 'Wt'
No pre-processing
```

```
Resampling: Cross-Validated (5 fold, repeated 5 times)
Summary of sample sizes: 4958, 4956, 4955, 4956, 4955,
4956, ...
Resampling results across tuning parameters:
  mtry  Accuracy   Kappa
   2    0.8034876  0.7151260
   7    0.8038741  0.7159579
  12    0.8021636  0.7136111

Accuracy was used to select the optimal model using the
 largest value.
The final value used for the model was mtry = 7.
```

In total 6,195 training samples were used for training. Here, 12 Landsat 5 TM bands are used as the predictors, while five land use/cover classes represent the response (target) variable. The best model had an mtry value of 7 with an overall accuracy of 0.804, which is relatively high (Fig. 4.24).

Next, let's display the variable importance using ***varImp()*** function (Fig. 4.25).

```
> rf_varImp <- varImp(rf_model)
> ggplot(rf_varImp, top = 10)
```

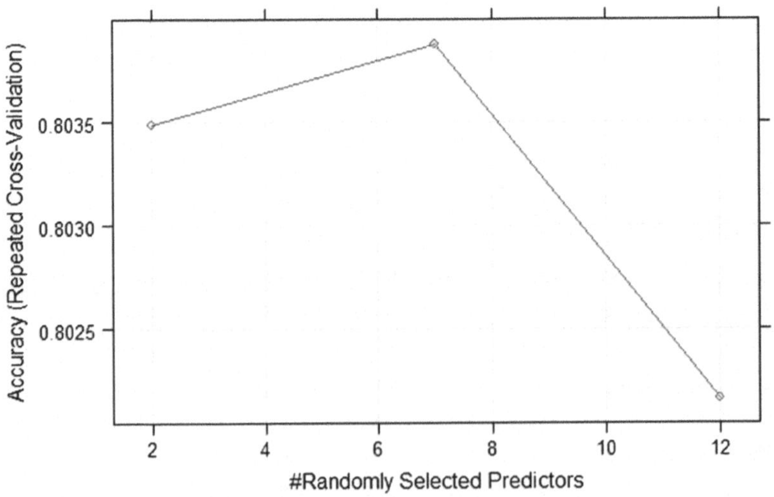

Fig. 4.24 Repeated cross-validation (based on overall accuracy) profile for the RF model

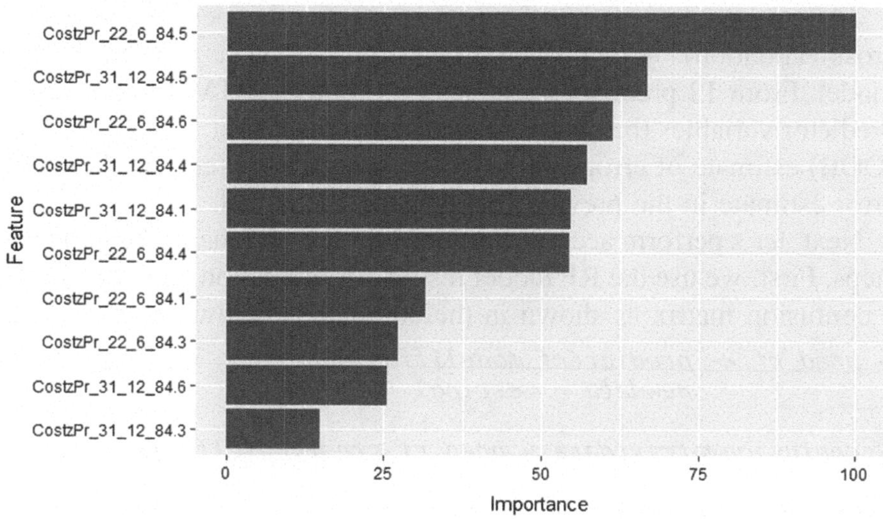

Fig. 4.25 Random forest model variable importance

Figure 4.25 shows the relative importance of the contribution of the 12 Landsat 5 TM bands. Band 5 (dry season) is the most informative predictor variable, followed by band 5 (wet season), and band 7 (dry season). As noted in Fig. 4.1, band 5 was better at separating the land use/cover classes. Next, let's check the parameters of the best model.

```
> rf_model$finalModel

Call:
 randomForest(x = x, y = y, mtry = param$mtry, proximit
y = TRUE,       fitBest = FALSE, returnData = TRUE)
               Type of random forest: classification
                     Number of trees: 500
No. of variables tried at each split: 7
        OOB estimate of  error rate: 18.43%
Confusion matrix:

     Agr   Br  GS Set Wt class.error
Agr 1585  336  30  17  1   0.1950229
Br   280 2072  32 135  0   0.1774514
GS    21   55 448  11  0   0.1626168
Set   31  129  36 912  0   0.1768953
Wt     3   20   3   2 36   0.4375000
```

The output shows a confusion matrix for the best model (after cross-validation). A total of 500 decision trees were used in the RF model. From 12 predictor variables (12 Landsat 5 TM bands), seven predictor variables (bands) were selected at each split. The out-of-bag (OOB) estimate of error rate is 18.4%, which is better than the OOB error estimate in the previous single date RF model.

Next, let's perform accuracy assessment as we done in the previous steps. First, we use the RF model results for prediction, and then build a confusion matrix as shown in the commands below.

```
> pred_rf <- predict(rf_model$finalModel,
              newdata = testing)

> confusionMatrix(data = pred_rf, testing$Cl1984)

Confusion Matrix and Statistics
          Reference
Prediction Agr  Br  GS Set  Wt
       Agr 401  56   8   8   0
       Br   81 536  17  29   1
       GS    6  10 103  11   2
       Set   4  27   3 229   0
       Wt    0   0   2   0  13


Overall Statistics
                    Accuracy : 0.8287
                      95% CI : (0.809, 0.8472)
         No Information Rate : 0.4066
         P-Value [Acc > NIR] : < 2.2e-16
                       Kappa : 0.7519
  Mcnemar's Test P-Value : NA

Statistics by Class:
                     Class: Agr Class: Br Class: GS Class: Set Class: Wt
Sensitivity              0.8150    0.8521   0.77444     0.8267  0.812500
Specificity              0.9318    0.8606   0.97949     0.9732  0.998694
Pos Pred Value           0.8478    0.8072   0.78030     0.8707  0.866667
Neg Pred Value           0.9153    0.8947   0.97880     0.9626  0.998042
Prevalence               0.3180    0.4066   0.08597     0.1791  0.010343
Detection Rate           0.2592    0.3465   0.06658     0.1480  0.008403
Detection Prevalence     0.3058    0.4292   0.08533     0.1700  0.009696
Balanced Accuracy        0.8734    0.8564   0.87696     0.9000  0.905597
```

The single date RF overall classification accuracy was 0.77, while the multidate RF overall classification accuracy was 0.83. This is quite a big improvement in accuracy. The RF classifier also performs better in terms of individual class performance. The producer's accuracy (sensitivity and specificity) and the user's accuracy (positive predicted values and negative predicted values) are much better for all land use/cover classes. While accuracy assessment results are relatively good, misclassification problems are still apparent. For example, 81 bareland pixels were misclassified as agriculture.

Step 11: Perform classification using the RF classifier

Now, let's perform land use/cover classification using the RF classifier.

```
> timeStart<- proc.time() # measure computation time
> LC_rf_84_Multdate <-predict(rasVars,rf_model)
> proc.time() - timeStart # user time and system time.
```

Step 12: Display the final land use/cover map

Let's display the final multidate land use/cover classification (Fig. 4.26).

```
> LC_ 84_Multdate <- gplot(LC_rf_84_Multdate) + geom_ra
ster(aes(fill = factor(value, labels=c("Agriculture", "
Bareland", "Green Spaces", "Urban", "Water")))) + scale
_fill_manual(values = c("yellow", "grey", "green3", "re
d", "blue3"), name= "Land Cover") + ggtitle("Random For
est Classification") +theme(plot.title = element_text(l
ineheight=.4, face="bold")) + coord_equal()
```

Finally, you can save the RF classification in tiff format using the command below.

```
>writeRaster(LC_rf_84_Multdate,
             "HreRF_Multidate_84.tif",
              type="raw",
              datatype='INT1U',
              index=1,
              na.rm=TRUE,
              progress="window",
              overwrite=TRUE)
```

The "HreRF_Multidate_84.tif" land use/cover map can be visualized in QGIS or other GIS software.

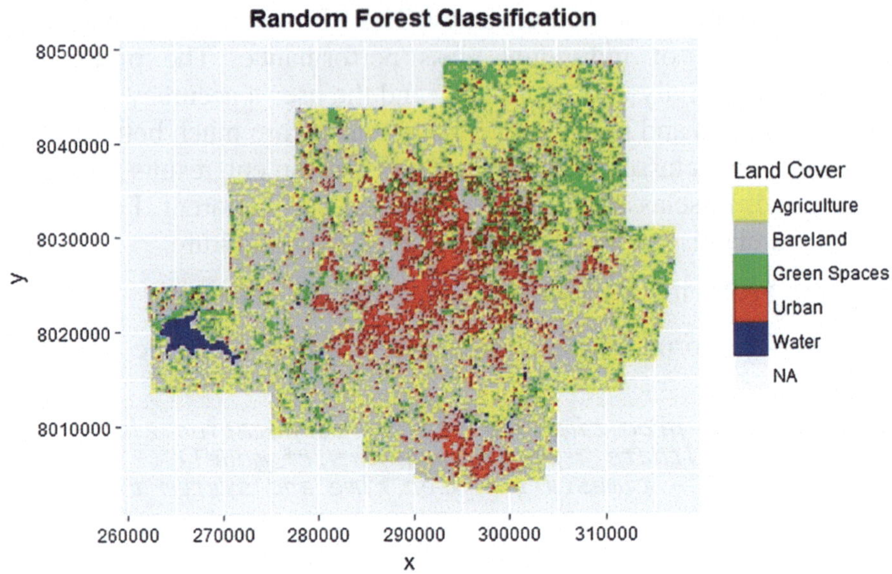

Fig. 4.26 Multidate land use/cover map

4.3.3 Summary for Tutorial 2

In tutorial 2, we classified multidate Landsat data set (which comprises Landsat 5 TM imagery acquired during the wet and dry seasons). The results indicate that the use of multidate datasets significantly improved land use/cover classification. While major improvements have been noted, we have also observed some misclassification problems.

4.4 Summary

In this chapter, we classified single date and multidate Landsat 5 TM imagery. First, we classified single date Landsat 5 TM imagery using the KNN, ANN, DT, radial basis SVM and RF classifiers. The results indicated that the RF classifier outperformed the KNN, ANN, DT and radial basis SVM classifiers. Based on that result, we used the RF classifier to classify multidate Landsat 5 TM imagery. In general, the RF classifier and multidate Landsat 5 TM imagery improved image classification. In Chap. 5, we are going to classify multiple data sets, which comprise multidate Landsat 5 TM imagery, and vegetation and texture indices. In addition, we are also going to check whether feature selection can improve classification accuracy.

4.5 Additional Exercises

i. Classify the multidate Landsat 5 TM imagery using the KNN and SVM classifiers.
ii. Compare the results from the RF, KNN and SVM classifiers.

References

Berberoglu S, Lloyd CD, Atkinson PM, Curran PJ (2000) The integration of spectral and textural information using neural networks for land cover mapping in the Mediterranean. Comput Geosci 26:385–396

Boser BE, Guyon IM, Vapnik VN (1992) A training algorithm for optimal margin classifiers. In: Proceedings of the fifth annual workshop on computational learning theory. ACM, pp 144–152. http://dl.acm.org/citation.cfm?id=130401. Accessed on 22 Feb 2014

Breiman L (1993) Classification and regression trees. Chapman & Hall, New York

Breiman L (2001) Random forests. Mach Learn 45:5–32

Breiman L, Friedman JH, Olshen RA, Stone CJ (1984) Classification and regression trees. Wadsworth International Group, Belmont, CA

Cortes C, Vapnik V (1995) Support-vector networks. Mach Learn 20 (3):273–297

Foody GM (2010) Assessing the accuracy of land cover change with imperfect ground reference data. Remote Sens Environ 114:2271–2285

Hsu CW, Chang CC, Lin CJ (2010) A practical guide to support vector classification. Department of Computer Science, National Taiwan University, Taipei, Taiwan, p. 16. http://www.csie.ntu.edu.tw/~cjlin/papers/guide/guide.pdf

Kavzoglu T, Mather PM (2003) The use of backpropagating artificial neural networks in land cover classification. Int J Remote Sens 10:4907–4938

Kuhn M (2008) Building predictive models in R using the caret package. J Stat Softw 28(5):1–26

Kuhn M, Johnson K (2016) Applied predictive modeling. Springer

Lu D, Weng Q (2007) A survey of image classification methods and techniques for improving classification performance. Int J Remote Sens 28:823–870

Mather PM (1999) Computer processing of remotely-sensed images-an introduction. Wiley, New York

Mather PM, Koch M (2011) Computer processing of remotely-sensed images: an introduction. Wiley-Blackwell, Chichester

Mellor A, Haywood A, Stone C, Jones S (2013) The performance of random forests in an operational setting for large area sclerophyll forest classification. Remote Sens 5(6):2838–2856

Miao X, Heaton JS, Zheng S, Charlet DA, Liu H (2012) Applying tree-based ensemble algorithms to the classification of ecological zones using multi-temporal multi-source remote-sensing data. Int J Remote Sens 33(6):1823–1849

Nemmour H, Chibani Y (2006) Multiple support vector machines for land cover change detection: an application for mapping urban extensions. ISPRS J Photogramm Remote Sens 61:125–133

Olden JD, Joy MK, Death RG (2004) An accurate comparison of methods for quantifying variable importance in artificial neural networks using simulated data. Ecol Modell 178:389–397

Olofsson P, Foody GM, Herold M, Stehman SV, Woodcock CE, Wulder MA (2014) Good practices for estimating area and assessing accuracy of land change. Remote Sens Environ 148:42–57

Pal M, Mather PM (2003) An assessment of the effectiveness of decision tree methods for land cover classification. Remote Sens Environ 86(4):554–565

Pal M, Mather PM (2005) Support vector machines for classification in remote sensing. Int J Remote Sens 26(5):1007–1011

Pontius RG Jr, Millones M (2011) Death to Kappa: birth of quantity disagreement and allocation disagreement for accuracy assessment. Int J Remote Sens 32:4407–4429

Quinlan JR (1986) Induction of decision trees. Mach Learn 1:81–106

Quinlan, JR (1993) C4.5: programs for machine learning. Morgan Kaufmann Publishers, San Mateo, California

Rodriguez-Galiano VF, Chica-Olmo M, Abarca-Hernandez F, Atkinson PM, Jeganathan C (2012) Random forest classification of Mediterranean land cover using multi-seasonal imagery and multi-seasonal texture. Remote Sens Environ 121:93–107

Sesnie SE, Gessler PE, Finegan B, Thessler S (2008) Integrating landsat TM and SRTM-DEM derived variables with decision trees for habitat classification and change detection in complex neotropical environments. Remote Sens Environ 112(5):2145–2159

Tso B, Mather PM (2001) Classification methods for remotely-sensed data. Taylor and Francis, New York

Walton JT (2008) Subpixel urban land cover estimation: comparing cubist, random forests, and support vector regression. Photogramm Eng Remote Sens 74(10):1213–1222

Yang X (2009) Artificial neural networks for urban modeling. In: Madden M (ed) Manual of geographic information systems. ASPRS, USA, pp 647–657

Zhang R, Ma J (2008) An improved SVM method P-SVM for classification of remotely-sensed data. Int J Remote Sens 29:6029–6036

Chapter 5
Improving Image Classification

Abstract Multidate imagery and satellite image derivatives such as vegetation and texture indices have been reported to improve image classification. However, the increase in additional predictor variables has also resulted in high data dimensionality and redundancy. Feature selection and extraction can be used to reduce high data dimensionality and minimize redundancy. The purpose of this chapter is to test whether feature selection can improve image classification. In this chapter, image classification will be performed using two different approaches. First, image classification is performed using the random forests (RF) classifier and multiple data sets (that consist of multidate Landsat 5 TM imagery, and vegetation and texture indices). Second, image classification is performed using the RF classifier with feature selection and multiple data sets. While the tutorial exercises indicate that feature selection did not improve image classification accuracy, it reduced the number of predictor variables.

Keywords High data dimensionality · Redundancy · Random forests (RF) classifier · Feature selection

---

Electronic supplementary material The online version of this chapter (https://doi.org/10.1007/978-981-13-8012-9_5) contains supplementary material, which is available to authorized users.

C. Kamusoko, *Remote Sensing Image Classification in R*, Springer Geography,
https://doi.org/10.1007/978-981-13-8012-9_5

5.1 Overview

Satellite image derivatives such as vegetation and texture indices as well as multidate images, multisensor images, and digital elevation model (DEM) have been reported to improve image classification. However, the increase in additional predictor variables results in high data dimensionality and redundancy. Feature selection and extraction can be used to reduce high data dimensionality and minimize redundancy (Dash and Liu 1997). A number of techniques such as principle component analysis have been used to reduce high data dimensionality and redundancy.

In this chapter, we are going to perform image classification using two different approaches. First, we are going to perform image classification using the RF classifier and multiple data sets, which comprises multidate Landsat 5 TM imagery, and vegetation and texture indices. Second, we are going to perform image classification using the RF classifier with feature selection and multiple data sets. Our main aim is to check whether the use of multiple data sets as well as feature selection improves image classification.

5.2 Feature Selection

5.2.1 Importance of Feature Selection

Machine learning classifiers are affected by high data dimensionality and noise, particularly if many predictor variables are used during classification (Yu and Liu 2004). Therefore, it is important to select optimal predictor variables because they: (i) enable the machine learning algorithm to train faster; and (ii) reduce model complexity. Filter and wrapper embedded methods are some of the most commonly used feature selection techniques (Kuhn and Johnson 2016).

Filter methods measure the relevance of features based on correlation with the response variable, while wrapper methods measure the usefulness of a subset of feature by actually training a model on it (Kuhn and Johnson 2016). Generally, filter methods evaluate each predictor in isolation before training the model or classifier (Kuhn and Johnson 2016). However, the wrapper methods evaluate multiple models in order to add or remove features from the subset, which eventually lead to the optimal model. Some common examples of wrapper methods are forward feature selection, backward feature elimination, recursive feature elimination (RFE). While filter methods are much faster compared to wrapper methods (since they do not involve training the models), they often fail to find the best subset of features, especially if there many highly correlated variables.

5.2.2 Recursive Feature Elimination (RFE)

In this chapter, we are going to use RFE in order to select optimal predictor variables. RFE is a greedy optimization algorithm that aims to find the best performing predictor variables. This technique repeatedly creates models and keeps aside the best or the worst performing feature at each iteration, and then constructs the next model with the remaining features until all the features are exhausted. RFE then ranks the features based on the order of their elimination.

5.3 Tutorial 1: Image Classification Using Multiple Data Sets

5.3.1 Objective

- Classify multiple data sets (that is, multidate Landsat 5 TM imagery, and vegetation and texture indices) using the RF classifier.

5.3.2 Procedure

In this exercise, we are going to classify multiple data sets (that is, multidate Landsat 5 TM imagery, and vegetation and texture indices) using the RF classifier. Following are the image classification steps.

Step 1: Load the necessary packages or libraries

First, load the following libraries

```
> library(raster)
> library(ggplot2)
> library(reshape)
> library(grid)
> library(gridExtra)
> library(RStoolbox)
> library(caret)
> library(rasterVis)
> library(corrplot)
> library(doParallel)
```

Step 2: Load raster data

Next, set up the working directory where all data sets are located.

```
> setwd("E:\\ToDatafolder")
```

Next, create a list of raster bands that will be used for classification.

```
> rasList <- list.files(getwd(),pattern="img$",
                 full.names=TRUE)
```

Combine or stack the raster layers, which you listed before.

```
> rasVars <- stack(rasList)
```

Once we have loaded and created a raster stack, the next step is to check the attributes and dimensions of the predictor variables.

```
> rasVars
class       : RasterStack
dimensions  : 1533, 1832, 2808456, 116  (nrow, ncol, ncell, nlayers)
resolution  : 30, 30  (x, y)
extent      : 262065, 317025, 8002765, 8048755  (xmin, xmax, ymin, ymax)
coord. ref.: +proj=utm +zone=36 +south +ellps=WGS84 +towgs84=0,0,0,-0,-0,-0,0 +units=m +no_
defs
names       : contrast_22_6_84_2, contrast_22_6_84_4, contrast_22_6_84_7, contrast_22_6_84_
b1, contrast_22_6_84_b3, contrast_22_6_84_b5, contrast_31_12_84_b1, contrast_31_12_84_b2, c
ontrast_31_12_84_b3, contrast_31_12_84_b4, contrast_31_12_84_b5, contrast_31_12_84_b7, corr
elation_22_6_
```

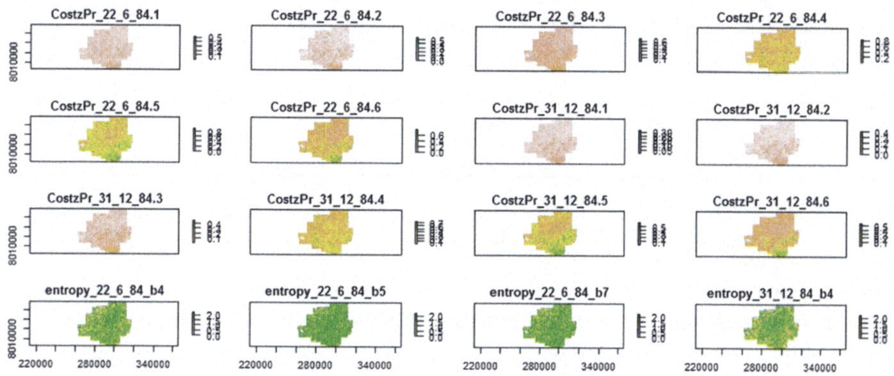

Fig. 5.1 Multidate landsat 5 TM imagery and other predictor variables

Next, display the predictor variables. However, this takes a bit of time since there are many predictor variables. Figure 5.1 shows a sample of the predictor variables.

```
> plot(rasVars)
```

Step 3: Load the training data

Load the training data set.

```
> ta_data <- readOGR(getwd(), "TA_1984")
OGR data source with driver: ESRI Shapefile
Source: "E:/Practical Machine Learning/ML_Woorkbook_Dat
asets/Data/ImageClassification/DS2_Landsat_Other_Variab
les/Dataset_MultiVariables", layer: "TA_1984"
with 7769 features
It has 1 fields
```

Check the attributes of the training data set.

```
> summary(ta_data)
```

```
Object of class SpatialPointsDataFrame
Coordinates:
                  min       max
coords.x1  262944.4  315205.6
coords.x2 8002868.9 8047944.6
Is projected: TRUE
proj4string :
[+proj=utm +zone=36 +south +datum=WGS84 +units=m +no_defs +el
lps=WGS84 +towgs84=0,0,0]
Number of points: 7769
Data attributes:
 cl1984
 Agr:2467
 Br :3169
 GS : 668
 Set:1385
 Wt :  80
```

Step 4: Use training points to extract reflectance values

Next, assign raster values to the training data set.

```
> ta<-as.data.frame(extract(rasVars, ta_data))
>ta_data@data=data.frame(ta_data@data,ta[match(rownames
(ta_data@data), rownames(ta)),])
```

Check the structure and attributes of the training data set. Note that the training data frame has 43 variables. Below is a sample of the training data set.

```
> str(ta_data@data)
```

```
'data.frame':      7769 obs. of  43 variables:
 $ cl1984               : Factor w/ 5 levels "Agr","Br","GS",..: 4
 4 4 4 4 4 4 4 4 4 ...
 $ CostzPr_22_6_84.1     : num  0.1195 0.1093 0.1059 0.0856 0.1161 .
 $ CostzPr_22_6_84.2     : num  0.1117 0.083 0.0687 0.0687 0.0973 ..
 $ CostzPr_22_6_84.3     : num  0.182 0.155 0.135 0.128 0.182 ...
 $ CostzPr_22_6_84.4     : num  0.284 0.234 0.209 0.192 0.284 ...
 $ CostzPr_22_6_84.5     : num  0.365 0.276 0.16 0.165 0.276 ...
 $ CostzPr_22_6_84.6     : num  0.338 0.265 0.127 0.16 0.233 ...
```

Next, check the NAs using the ***complete.cases()*** function.

```
> complete.cases(ta_data@data)
 [1]  TRUE  TRUE  TRUE  TRUE  TRUE  TRUE  TRUE  TRUE  TRUE  TRUE  TRUE  TRUE  TRUE  TRUE
TRUE  TRUE  TRUE  TRUE  TRUE  TRUE
[21]  TRUE  TRUE  TRUE  TRUE  TRUE  TRUE FALSE  TRUE  TRUE  TRUE  TRUE  TRUE  TRUE  TRUE
TRUE FALSE  TRUE FALSE  TRUE  TRUE
[41] FALSE  TRUE FALSE  TRUE  TRUE  TRUE  TRUE FALSE FALSE FALSE  TRUE  TRUE FALSE FALSE
FALSE FALSE FALSE FALSE  TRUE  TRUE
[61] FALSE  TRUE FALSE FALSE FALSE  TRUE FALSE  TRUE  TRUE  TRUE  TRUE FALSE  TRUE FALSE
TRUE  TRUE  TRUE FALSE  TRUE  TRUE
[81]  TRUE  TRUE FALSE  TRUE  TRUE  TRUE FALSE FALSE FALSE FALSE  TRUE FALSE FALSE FALSE
TRUE  TRUE  TRUE FALSE FALSE  TRUE
[ reached getOption("max.print") -- omitted 6769 entries ]
```

The results show that there are some NAs. Run the ***dim()*** function.

```
> dim(ta_data@data)
[1] 7769   43
```

The output shows that the training dataset has 7,769 points with 43 variables (one response variable and 42 predictor variables).

Next, remove the NAs from the data frame in order to avoid problems during model training.

```
> ta_data@data <- na.omit(ta_data@data)
> complete.cases(ta_data@data)
 [1] TRUE TRUE TRUE TRUE TRUE TRUE TRUE TRUE TRUE TRUE TRUE TRUE TRUE TRUE TRUE TRUE TRUE T
RUE TRUE TRUE TRUE TRUE TRUE TRUE
 [25] TRUE TRUE TRUE TRUE TRUE TRUE TRUE TRUE TRUE TRUE TRUE TRUE TRUE TRUE TRUE TRUE TRUE
TRUE TRUE TRUE TRUE TRUE TRUE TRUE
[49] TRUE TRUE TRUE TRUE TRUE TRUE TRUE TRUE TRUE TRUE TRUE TRUE TRUE TRUE TRUE TRUE TRUE T
RUE TRUE TRUE TRUE TRUE TRUE TRUE
[73] TRUE TRUE TRUE TRUE TRUE TRUE TRUE TRUE TRUE TRUE TRUE TRUE TRUE TRUE TRUE TRUE TRUE T
RUE TRUE TRUE TRUE TRUE TRUE TRUE
[ reached getOption("max.print") -- omitted 3099 entries ]
```

The output shows that NAs have been removed. Next, check the training data again using the ***dim()*** function.

```
> dim(ta_data@data)
[1] 7732   43
```

Note that we removed a small proportion of the training data set. Therefore, the remaining data set is still good enough to serve our purpose.

Please note that, RF classifier does not handle missing values in predictors. Therefore, training data with missing values can be either removed or imputed (fill in missing values) with the median (for numerical values) or mode (for categorical values). There are many techniques for imputing missing data. While it is not recommended to remove training points in case you do not have enough training, imputing data is computer intensive. For the purpose of this tutorial, we are going to remove the NAs since we have a small proportion of missing values.

Step 5: Prepare training and test data sets

First, let's set a pre-defined value as we done before.

```
> hre_seed<- 27
> set.seed(hre_seed)
```

Next, split the dataset into training and test sets.

```
> inTraining<- createDataPartition(ta_data@data$Cl1984,
  p = .80, list = FALSE)
> training<- ta_data@data[ inTraining,]
> testing  <- ta_data@data[-inTraining,]
```

Check the dimensions and size of the training and testing sets using the ***dim()*** function.

```
> dim(training)
[1] 6187   43

> dim(testing)
[1] 1545   43
```

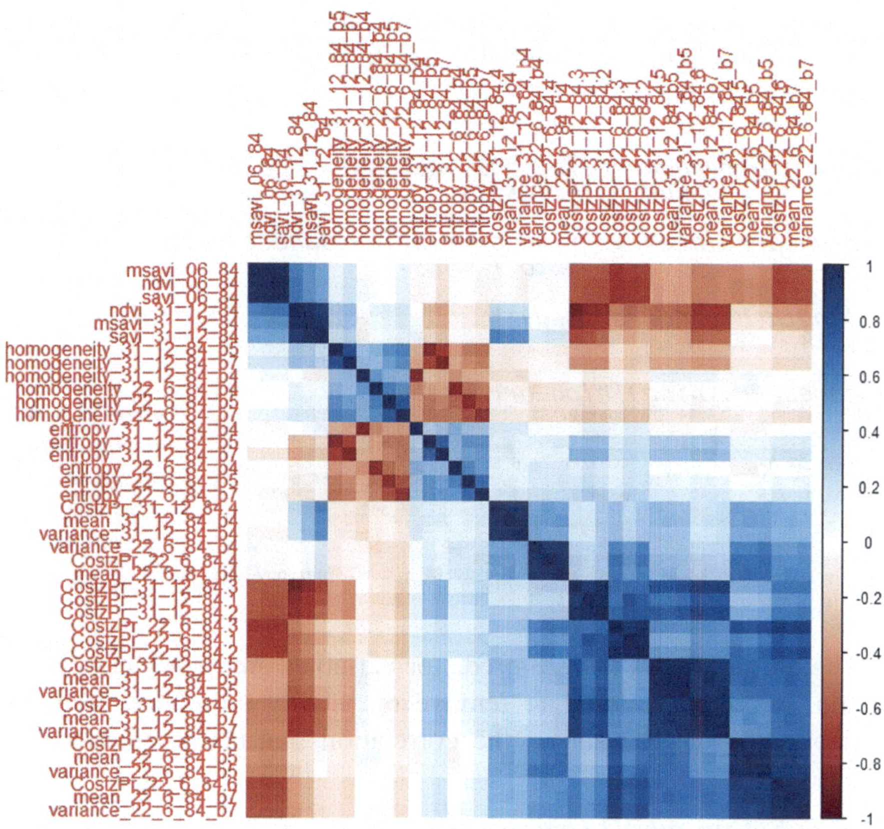

Fig. 5.2 Correlation matrix for all the predictor variables

The output shows that the training data has 6,187 training points and 43 variables (one response and 42 predictor variables), while the testing set has 1,545 testing points.

Let's check the correlations of all the predictor variables using the ***cor()*** and ***corrplot()*** functions.

```
> bandCorrelations <- cor(training[, 2:43])

> dim(bandCorrelations)
[1] 42 42
```

We are going to plot the correlation matrix (Fig. 5.2), and then check the predictor variables that are highly correlated based on a cutoff point of 0.8.

```
> corrplot(bandCorrelations, method = "color",
           order = "hclust")

> highCorrelated <- findCorrelation(bandCorrelations,
                    cutoff = .8)

> length(highCorrelated)
[1] 28

> names(training)[highCorrelated]
 [1] "CostzPr_31_12_84.1" "CostzPr_31_12_84.2"       "CostzPr_22_6_84.1" "CostzPr_22_6_84.2"
"variance_22_6_84_b5"
 [6]  "CostzPr_22_6_84.5"                 "mean_22_6_84_b5"                 "CostzPr_31_12_84.5"
"mean_31_12_84_b5"          "CostzPr_22_6_84.6"
[11]  "CostzPr_31_12_84.4"                 "variance_31_12_84_b4"          "mean_31_12_84_b4"
"variance_22_6_84_b4"       "mean_22_6_84_b4"
[16]  "ndvi_06_84"                         "msavi_06_84"                       "ndvi_31_12_84"
"mean_31_12_84_b7"          "homogeneity_31_12_84_b7"
[21]  "homogeneity_31_12_84_b5"  "savi_31_12_84"                      "CostzPr_31_12_84.3"
"homogeneity_31_12_84_b4" "variance_22_6_84_b7"
[26] "entropy_22_6_84_b4"        "entropy_31_12_84_b7"     "entropy_22_6_84_b7"
```

The output shows that 28 predictor variables are highly correlated. Again, we are not going to remove the highly correlated predictor variables. Figure 5.2 shows the correlation matrix for all predictor variables.

Step 6: Set-up model tuning parameters

Next, use "set.seed" to set up the pre-defined values as we have done previously.

```
set.seed(hre_seed)
```

After that, set up the model tuning parameters.

```
> fitControl<- trainControl(
                  method = "repeatedcv",
                  number = 5,
                  repeats = 5)
```

Step 7: Train the Random Forest (RF) Classifier

We are going to start training the RF model as shown below.

```
> cl <- makePSOCKcluster(5)
> registerDoParallel(cl)
> set.seed(hre_seed)

> timeStart<- proc.time()

> rf_model<-train(Cl1984~.,data=training, method="rf",
                  trControl=fitControl,
                  prox=TRUE,
                  fitBest = FALSE,
                  returnData = TRUE)

> proc.time() - timeStart # user time and system time.

      User        System   Elapsed
      44.55        0.52    1039.05
```

It took approximately 17 min to train the RF model.

```
> stopCluster(cl) # stop when you are done:
```

Next, let's check the performance of the RF classifier. Again we use the ***print()*** and ***plot()*** functions as shown in the commands below.

```
> print(rf_model)
> plot (rf_model)

Random Forest
6187 samples
  42 predictor
   5 classes: 'Agr', 'Br', 'GS', 'Set', 'Wt'

No pre-processing
Resampling: Cross-Validated (5 fold, repeated 5 times)
Summary of sample sizes: 4950, 4949, 4950, 4948, 4951,
4950, ...
Resampling results across tuning parameters:
  mtry  Accuracy   Kappa
   2    0.8386631  0.7658309
  22    0.8404744  0.7686974
  42    0.8372093  0.7640465

Accuracy was used to select the optimal model using the
largest value.
The final value used for the model was mtry = 22.
```

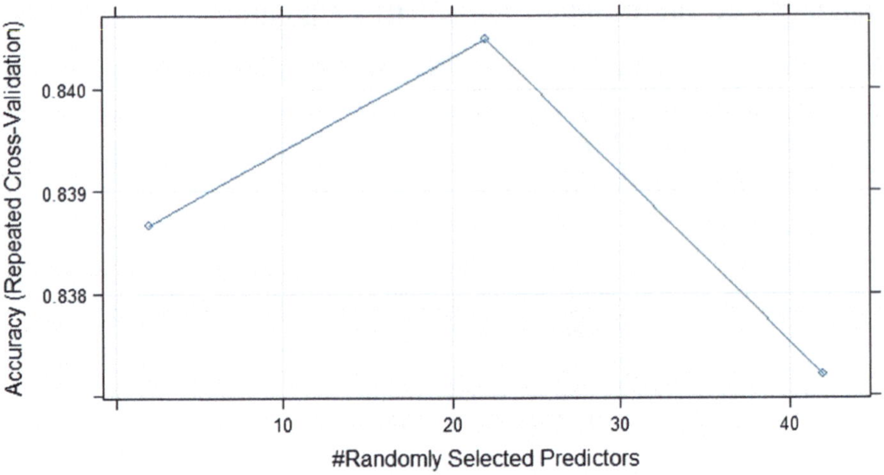

Fig. 5.3 Repeated cross-validation (based on overall accuracy) profile for the RF model

The results show that 6,187 training samples were used for training. Five land use/cover classes represent the response variable, and there are 42 predictor variables. The best model had an mtry value of 22 with an overall accuracy of 84%, which is relatively high (Fig. 5.3).

Next, check the parameters of the best model.

```
Call:
 randomForest(x = x, y = y, mtry = param$mtry,
proximity = TRUE,        fitBest = FALSE, returnData =
TRUE)
               Type of random forest: classification
                     Number of trees: 500
No. of variables tried at each split: 22

        OOB estimate of  error rate: 14.85%
Confusion matrix:
      Agr   Br  GS Set Wt class.error
Agr 1620  311  18  14  0   0.1747326
Br   191 2196  29 101  0   0.1275328
GS    20   49 451  15  0   0.1570093
Set   14  109  26 959  0   0.1344765
Wt     2   16   3   1 42   0.3437500
```

The output shows a confusion matrix for the best model (after cross-validation). A total of 500 decision trees were used in the RF model. The out-of-bag (OOB) estimate of error rate is 14.85%. The

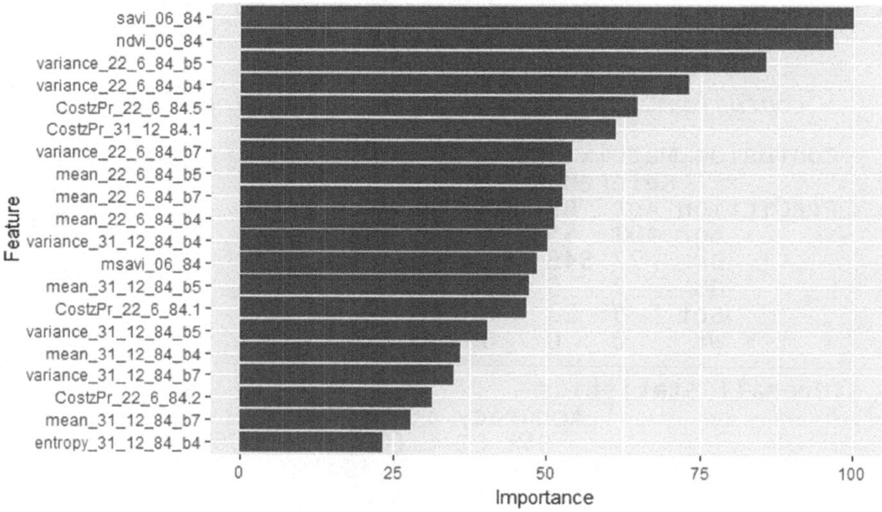

Fig. 5.4 Random forest variable importance

result is relatively better compared to the previous RF model in Chap. 4.

Next, let's us check the variable importance using ***varImp()*** function (Fig. 5.4).

```
> rf_varImp <- varImp(rf_model, compete = FALSE)
> plot(rf_varImp, top = 20)
```

Figure 5.4 shows the relative importance of the contribution of the top 20 predictor variables. The top 5 predictors are from the dry season Landsat 5 TM imagery (acquired in June), followed by Landsat 5 TM band 1 from the wet season (acquired in December). In particular, NDVI and SAVI, bands 4 and 5 variance, and Landsat 5 TM band 5 (all dry season) are the most informative predictor variables. As observed in Fig. 4.1, band 5 is better at separating land use/cover in the test site.

Next, let's perform accuracy assessment as we done in the previous steps. First, we use the RF model results for prediction, and then build a confusion matrix as shown in the commands below.

```
> pred_rf<- predict(rf_model$finalModel,
              newdata = testing)

> confusionMatrix(data = pred_rf, testing$Cl1984)

Confusion Matrix and Statistics
          Reference
Prediction Agr  Br  GS Set  Wt
       Agr 401  55   7   7   2
       Br   77 546   7  30   5
       GS    7   5 119   3   2
       Set   5  23   0 237   0
       Wt    0   0   0   0   7

 Overall Statistics
                Accuracy : 0.8479
                  95% CI : (0.829, 0.8655)
     No Information Rate : 0.4071
     P-Value [Acc > NIR] : < 2.2e-16
                   Kappa : 0.7794
  Mcnemar's Test P-Value : NA

 Statistics by Class:
                     Class: Agr Class: Br Class: GS Class: Set Class: Wt
Sensitivity              0.8184    0.8680   0.89474     0.8556  0.437500
Specificity              0.9327    0.8701   0.98796     0.9779  1.000000
Pos Pred Value           0.8496    0.8211   0.87500     0.8943  1.000000
Neg Pred Value           0.9171    0.9057   0.99006     0.9688  0.994148
Prevalence               0.3172    0.4071   0.08608     0.1793  0.010356
Detection Rate           0.2595    0.3534   0.07702     0.1534  0.004531
Detection Prevalence     0.3055    0.4304   0.08803     0.1715  0.004531
Balanced Accuracy        0.8755    0.8691   0.94135     0.9168  0.718750
```

There is a slight increase in overall classification accuracy compared to the previous multidate RF model in Chap. 4. The current RF model has an overall accuracy of 85% versus 83% from the multidate image classification.

Step 8: Perform classification using the RF classifier

Now, let's perform land use/cover classification using the RF classifier.

```
> timeStart <- proc.time() # measure computation time

> LC_rf_84 <-predict(rasVars,rf_model)

> proc.time() - timeStart # user time and system time.

user   system elapsed
132.23       7.39      140.04
```

Step 9: Display final land use/cover

Let's display the final land use/cover map (Fig. 5.5).

```
> LC_rf_84_MultiVariables1 <- ggplot(LC_rf_84) +
      geom_raster(aes(fill = factor(value,
      labels=c("Agriculture", "Bareland", "Green
      Spaces", "Urban", "Water")))) +
      scale_fill_manual(values = c("yellow", "grey",
      "green3", "red", "blue3"), name= "Land Cover") +
      ggtitle("Random Forest Classification")
     +theme(plot.title = element_text(lineheight=.4,
     face="bold")) + coord_equal()

> LC_rf_84_MultiVariables1
```

Finally, you can save the land cover map if you want to visualize it in QGIS or other GIS software packages.

```
>writeRaster(LC_rf_84,
             "RF84_MultiVariables1.tif",
              type="raw",
              datatype='INT1U',
              index=1,
              na.rm=TRUE,
              progress="window",
              overwrite=TRUE)
```

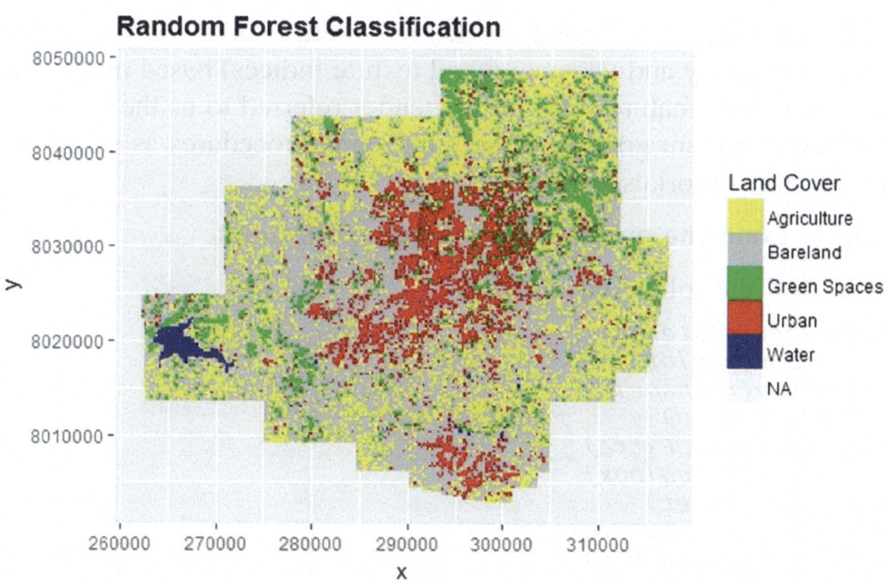

Fig. 5.5 Land cover map classified using the RF classifier and multiple datasets (without feature selection)

5.3.3 *Summary for Tutorial 1*

In tutorial 1, we observed that the multiple data sets (which comprises multidate Landsat 5 TM imagery, vegetation and texture indices) slightly improved land use/cover classification. However, we also observed high between predictor correlations. Next, let's try to optimize the RF model using feature selection, and see if we can improve classification.

5.4 Tutorial 2: Image Classification Using Multiple Data Sets with Feature Selection

5.4.1 *Objective*

- Classify multiple data sets (that is, multidate Landsat 5 TM imagery, and vegetation and texture indices) using the RFE-RF classifier (with feature selection).

5.4.2 *Procedure*

In this tutorial, we are going to classify multiple data sets (multidate Landsat imagery and vegetation and texture indices) based on the RF algorithm with feature selection (hereafter referred to as the RFE-RF classifier). We are going to follow the same procedures as shown in the previous tutorials.

Step 1: Load the necessary packages or libraries

First, load the following libraries.

```
> library(raster)
> library(ggplot2)
> library(reshape)
> library(grid)
> library(gridExtra)
> library(RStoolbox)
> library(caret)
> library(randomForest)
> library(rasterVis)
> library(corrplot)
> library(doParallel)
```

Step 2: Set up your working directory and load raster data

Next, set up the working directory.

```
> setwd("E:\\ToDataFolder\\Dataset_MultiVariables")
```

Create a list of predictor variables, which will be used for classification.

```
> rasList <- list.files(getwd(),pattern="img$",
               full.names=TRUE)
```

Combine or stack the layers, which you listed before.

```
> rasVars <- stack(rasList)
```

Next, check the attributes and dimensions of all the predictor variables.

```
> rasVars
class       : RasterStack
dimensions  : 1533, 1832, 2808456, 116  (nrow, ncol, ncell, nlayers)
resolution  : 30, 30  (x, y)
extent      : 262065, 317025, 8002765, 8048755  (xmin, xmax, ymin, ymax)
coord. ref. : +proj=utm +zone=36 +south +ellps=WGS84 +towgs84=0,0,0,-0,-0,-0,0 +units=m +no
_defs
names       : contrast_22_6_84_2, contrast_22_6_84_4, contrast_22_6_84_7, contrast_22_6_84_
b1, contrast_22_6_84_b3, contrast_22_6_84_b5, contrast_31_12_84_b1, contrast_31_12_84_b2, c
ontrast_31_12_84_b3, contrast_31_12_84_b4, contrast_31_12_84_b5, contrast_31_12_84_b7, corr
elation_22_6_
```

Step 3: Load the training data

Next, load training data.

```
> ta_data <- readOGR(getwd(), "TA_1984")
OGR data source with driver: ESRI Shapefile
Source: "E:/Practical Machine Learn-
ing/ML_Woorkbook_Datasets/Data/ImageClassification/DS2_
Landsat_Other_Variables/Dataset_MultiVariables", layer:
"TA_1984" with 7769 features
It has 1 fields
```

Step 4: Use training points to extract reflectance values

Then assign raster values to training data.

```
>ta<-as.data.frame(extract(rasVars, ta_data))
>ta_data@data=data.frame(ta_data@data,ta[match(rownames
(ta_data@data), rownames(ta)),])
```

Check the structure and attributes of the training data.

```
> str(ta_data@data)
```

```
'data.frame':      7769 obs. of  117 variables:
 $ Cl1984                   : Factor w/ 5 levels "Agr","Br","GS",..:
 4 4 4 4 4 4 4 4 4 4 ...
 $ contrast_22_6_84_2       : num  5.44 5.78 5.78 11.67 3.22 ...
 $ contrast_22_6_84_4       : num  5.22 5.33 10.44 14.44 3.11 ...
 $ contrast_22_6_84_7       : num  25.11 15.11 8.44 13 2.67 ...
 $ contrast_22_6_84_b1      : num  3.22 2.33 7.11 16.44 3.67 ...
 $ contrast_22_6_84_b3      : num  6.11 6.22 6.89 13.67 2.56 ...
 $ contrast_22_6_84_b5      : num  24.56 11.22 10.33 20.56 3.67 ...
 $ contrast_31_12_84_b1     : num  6.67 10.33 22.33 24.11 11.67 ...
 $ contrast_31_12_84_b2     : num  3.89 6.44 13.56 12.22 5.33 ...
 $ contrast_31_12_84_b3     : num  3.67 5 11.89 8.11 2.56 ...
```

We know that there are some NAs in our training data frame as we observed in Sect. 5.4. Next, remove the NAs from the data frame.

```
> ta_data@data <- na.omit(ta_data@data)
> complete.cases(ta_data@data)
```

```
[1] TRUE TRUE TRUE TRUE TRUE TRUE TRUE TRUE TRUE TRUE TRUE TRUE TRUE TRUE TRUE TRUE TRUE
TRUE TRUE TRUE TRUE TRUE TRUE TRUE
 [25] TRUE TRUE TRUE TRUE TRUE TRUE TRUE TRUE TRUE TRUE TRUE TRUE TRUE TRUE TRUE TRUE TRUE
TRUE TRUE TRUE TRUE TRUE TRUE TRUE
 [49] TRUE TRUE TRUE TRUE TRUE TRUE TRUE TRUE TRUE TRUE TRUE TRUE TRUE TRUE TRUE TRUE TRUE
TRUE TRUE TRUE TRUE TRUE TRUE TRUE
 [73] TRUE TRUE TRUE TRUE TRUE TRUE TRUE TRUE TRUE TRUE TRUE TRUE TRUE TRUE TRUE TRUE TRUE
TRUE TRUE TRUE TRUE TRUE TRUE TRUE
 [ reached getOption("max.print") -- omitted 3099 entries ]
```

Step 5: Prepare training and test data sets

Set a pre-defined value.

```
> hre_seed<- 27
> set.seed(hre_seed)
```

Split the training data set into training and test sets.

```
> inTraining <- createDataPartition(ta_data@data$Cl1984,
                   p = .80, list = FALSE)
> training<- ta_data@data[ inTraining,]
> testing  <- ta_data@data[-inTraining,]

> dim(training)
[1] 6187   43

> dim(testing)
[1] 1545   43
```

Step 6: Train the RFE-RF model

We are going to train the RFE-RF model using automatic feature selection. First, we are going to create a "control" object using the ***rfeControl()*** function as specified in the command below.

```
> control <- rfeControl(functions=rfFuncs,
                        method="repeatedcv",
                        number=5)
```

Next, run the RFE algorithm. Notice that the model specification is different here. Furthermore, this model takes some time to run. So be patient!!!

```
> cl <- makePSOCKcluster(5)
> registerDoParallel(cl)
> set.seed(hre_seed)

> timeStart<- proc.time()

> rfe_model <- rfe(training[,2:43],
                   training[,1],
                   sizes=c(2:43),
                   method="rf",
                   rfeControl=control,
                   prox=TRUE,
                   fitBest = FALSE,
                   returnData = TRUE)
> proc.time() - timeStart
  User   System        Elapsed
 43.72    0.67         908.46
```

It took approximately 26 min to run the RF model. Next, print and check the model results.

```
> print(rfe_model)

Outer resampling method: Cross-Validated (10 fold, repe
ated 1 times)

Resampling performance over subset size:
 Variables Accuracy  Kappa AccuracySD KappaSD Selected
         2   0.6675 0.5191   0.012846 0.01845
         3   0.6890 0.5484   0.024414 0.03579
         4   0.7154 0.5872   0.027100 0.03962
                     .
                     .
        27   0.8507 0.7835   0.015149 0.02176
        28   0.8536 0.7879   0.014813 0.02119
        29   0.8491 0.7812   0.014280 0.02024
        30   0.8484 0.7803   0.015216 0.02176
        31   0.8500 0.7827   0.015722 0.02252
                     .
        40   0.8518 0.7853   0.014626 0.02113
        41   0.8498 0.7825   0.015014 0.02175
        42   0.8492 0.7816   0.012884 0.01853
The top 5 variables (out of 28):
   variance_22_6_84_b5, variance_22_6_84_b4, mean_22_6_
84_b4, mean_22_6_84_b5, variance_31_12_84_b4
```

The RFE algorithm selected “variance_22_6_84_b5”, “variance_22_6_84_b4”, “mean_22_6_84_b4”, “mean_22_6_84_b5”, and “variance_31_12_84_b4” as the top five variables.

Next, list the selected features. In total 28 predictor variables were selected from 42. Next, check the selected predictor variables using the command below.

```
> predictors(rf_model)

 [1] "variance_22_6_84_b5" "variance_22_6_84_b4" "mean_22_6_84_b4"
mean_22_6_84_b5" "variance_31_12_84_b4" "variance_31_12_84_b5"  "mea
n_31_12_84_b5" "variance_22_6_84_b7""mean_31_12_84_b4" "CostzPr_31_1
2_84.1" "CostzPr_22_6_84.5" "mean_22_6_84_b7" "variance_31_12_84_b7"
"ndvi_06_84" "savi_06_84" "msavi_06_84" "mean_31_12_84_b7" "CostzPr_
22_6_84.1" "entropy_31_12_84_b4" "homogeneity_31_12_84_b4" "homogene
ity_22_6_84_b4"  "CostzPr_31_12_84.5" "savi_31_12_84" "CostzPr_22_6_
84.4" "CostzPr_31_12_84.4" "homogeneity_31_12_84_b5" "CostzPr_22_6_8
4.2" "CostzPr_22_6_84.6"
```

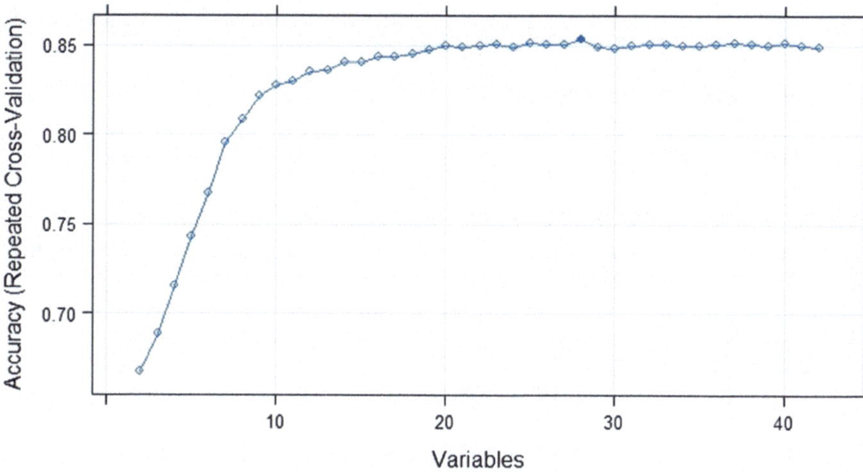

Fig. 5.6 Plot the RFE model variable selection

Now, let's plot the RFE model accuracy based on cross validation (Fig. 5.6).

```
plot(rf_model, type=c("g", "o"))
```

Figure 5.6 shows that 28 predictor variables were selected. The RFE model accuracy is 85%. It should be noted that we can use the RFE model for prediction. However, in this workbook, we are only using the RFE to select the relevant predictor variables. The selected predictor variables are going to be used for the final RF model training.

Step 7: Train the RF model using the selected predictor variables

Let's specify the 28 top predictor variables from the RFE model.

```
> (f <- as.formula(paste("Cl1984", paste(rf_model$optVariables, col-
lapse=" + "), sep=" ~ ")))

Cl1984 ~ variance_22_6_84_b5 + variance_22_6_84_b4 + mean_22_6_84_b4 +
    mean_22_6_84_b5 + variance_31_12_84_b4 + variance_31_12_84_b5 +
    mean_31_12_84_b5 + variance_22_6_84_b7 + mean_31_12_84_b4 +
    CostzPr_31_12_84.1 + CostzPr_22_6_84.5 + mean_22_6_84_b7 +
    variance_31_12_84_b7 + ndvi_06_84 + savi_06_84 + msavi_06_84 +
    mean_31_12_84_b7 + CostzPr_22_6_84.1 + entropy_31_12_84_b4 +
    homogeneity_31_12_84_b4 + homogeneity_22_6_84_b4 + CostzPr_31_12_84.5 +
    savi_31_12_84 + CostzPr_22_6_84.4 + CostzPr_31_12_84.4 +
    homogeneity_31_12_84_b5 + CostzPr_22_6_84.2 + CostzPr_22_6_84.6
```

Next, set up the model tuning parameters as we have done in tutorial 1.

```
> fitControl<- trainControl(
      method = "repeatedcv",
     number = 5,
     repeats = 5)

> set.seed(hre_seed)
```

After that, we can set up the RF model and execute the commands below.

```
> timeStart<- proc.time()
> opt_rfmodel<-train(f,
                    data=training,
                    method="rf",
                    trControl=fitControl,
                    prox=TRUE,
                    fitBest = FALSE,
                    returnData = TRUE)
> proc.time() - timeStart
   User    System        Elapsed
   35.52         0.53      855.67

> stopCluster(cl)
```

It took approximately 14 min to run the RF model with the optimal predictor variables.

Next, let's check the performance of the RF classifier. Again we use the ***print()*** and ***plot()*** functions as shown in the commands below.

```
> print(opt_rfmodel)
> plot(opt_rfmodel)

                    6187 samples
                     28 predictor
        5 classes: 'Agr', 'Br', 'GS', 'Set', 'Wt'

                  No pre-processing
 Resampling: Cross-Validated (5 fold, repeated 5 times)
 Summary of sample sizes: 4950, 4949, 4950, 4948, 4951,
                     4950, ...
      Resampling results across tuning parameters:
                  mtry  Accuracy   Kappa
                   2    0.8419927  0.7709048
                  15    0.8400851  0.7682892
                  28    0.8359803  0.7623514

Accuracy was used to select the optimal model using the
largest value.
The final value used for the model was mtry = 2.
```

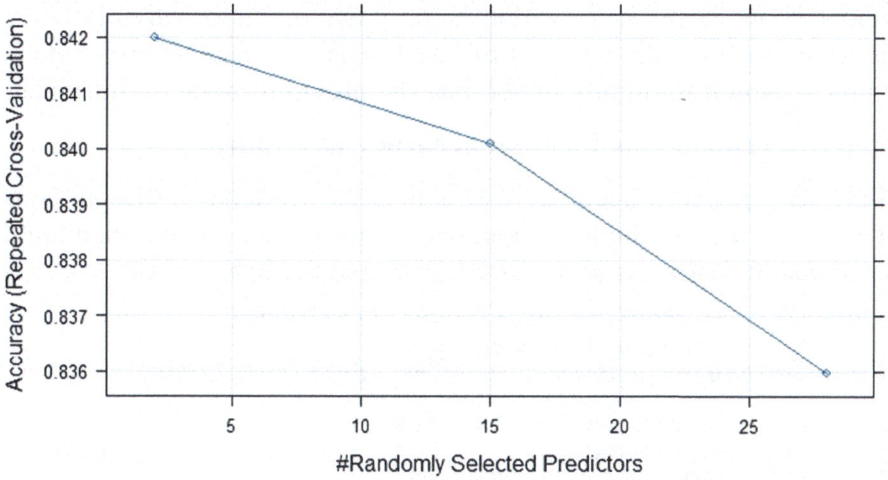

Fig. 5.7 Repeated cross-validation (based on overall accuracy) profile for the RF model

The output shows that 6,187 training samples were used for training. Here, there are five land use/cover classes that represent the response variable, and 28 optimal predictor variables. The best model had an mtry value of 2 with an overall accuracy of 84% (Fig. 5.7).

Next, check the best model results.

```
> opt_rfmodel$finalModel

Call:
 randomForest(x = x, y = y, mtry = param$mtry, proximit
y = TRUE,       fitBest = FALSE, returnData = TRUE)
               Type of random forest: classification
                     Number of trees: 500
No. of variables tried at each split: 2

        OOB estimate of  error rate: 14.64%
Confusion matrix:
     Agr   Br  GS Set Wt class.error
Agr 1611  319  23  10  0   0.1793174
Br   189 2208  27  91  2   0.1227652
GS    16   50 454  15  0   0.1514019
Set   12  102  27 967  0   0.1272563
Wt     2   18   1   2 41   0.3593750
```

The output shows a confusion matrix for the best model (after cross-validation). A total of 500 decision trees were used in the RF model. From 28 predictor variables, only two predictor variables were selected at each split. The out-of-bag (OOB) estimate of error rate is 14.64%, which is slightly better than the previous model in tutorial 1.

Step 8: Perform classification accuracy assessment

Next, let's perform accuracy assessment as we done in the previous steps. First, we use the RF model results for prediction, and then build a confusion matrix as shown in the commands below.

```
> pred_rf <- predict(opt_rfmodel$finalModel,
                newdata = testing)
> confusionMatrix(data = pred_rf, testing$Cl1984)
```

```
Confusion Matrix and Statistics
          Reference
Prediction Agr  Br   GS Set  Wt
       Agr 401  63    7   4   0
       Br   77 542    7  28   8
       GS    8   5  119   6   2
       Set   4  19    0 239   0
       Wt    0   0    0   0   6

 Overall Statistics
                    Accuracy : 0.846
                      95% CI : (0.827, 0.8636)
         No Information Rate : 0.4071
         P-Value [Acc > NIR] : < 2.2e-16
                       Kappa : 0.7767
  Mcnemar's Test P-Value : NA

Statistics by Class:
                     Class: Agr Class: Br Class: GS Class: Set Class: Wt
Sensitivity              0.8184    0.8617   0.89474     0.8628  0.375000
Specificity              0.9299    0.8690   0.98513     0.9819  1.000000
Pos Pred Value           0.8442    0.8187   0.85000     0.9122  1.000000
Neg Pred Value           0.9168    0.9015   0.99004     0.9704  0.993502
Prevalence               0.3172    0.4071   0.08608     0.1793  0.010356
Detection Rate           0.2595    0.3508   0.07702     0.1547  0.003883
Detection Prevalence     0.3074    0.4285   0.09061     0.1696  0.003883
Balanced Accuracy        0.8741    0.8653   0.93993     0.9223  0.687500
```

The overall classification accuracy for the RF-RFE (with feature selection) model is 84.6%, while the overall accuracy for the simple RF model (without feature selection) is 84.8%. This shows that feature selection based on the RFE model did not improve land use/cover classification. However, the RFE managed to reduce the predictor variables and thus reduced computation time for the final RF model.

Step 9: Perform classification using the RF classifier

Now, let's perform land use/cover classification using the RF classifier.

```
> timeStart<- proc.time() # measure computation time
> LC_rf_84 <-predict(rasVars,opt_rfmodel)
> proc.time() - timeStart # user time and system time.

User        System      Elapsed
136.49        7.12      145.55
```

Step 10: Display final land use/cover classifications

Let's display all the final land use/cover classification (Fig. 5.8).

```
> LC_rf_84_MultiVariables2 <- gplot(LC_rf_84) +
        geom_raster(aes(fill = factor(value,
        labels=c("Agriculture", "Bareland", "Green
        Spaces", "Urban", "Water")))) +
        scale_fill_manual(values = c("yellow", "grey",
        "green3", "red", "blue3"), name= "Land Cover") +
        ggtitle("Random Forest Classification")
       +theme(plot.title = element_text(lineheight=.4,
       face="bold")) + coord_equal()
       > LC_rf_84_MultiVariables2
```

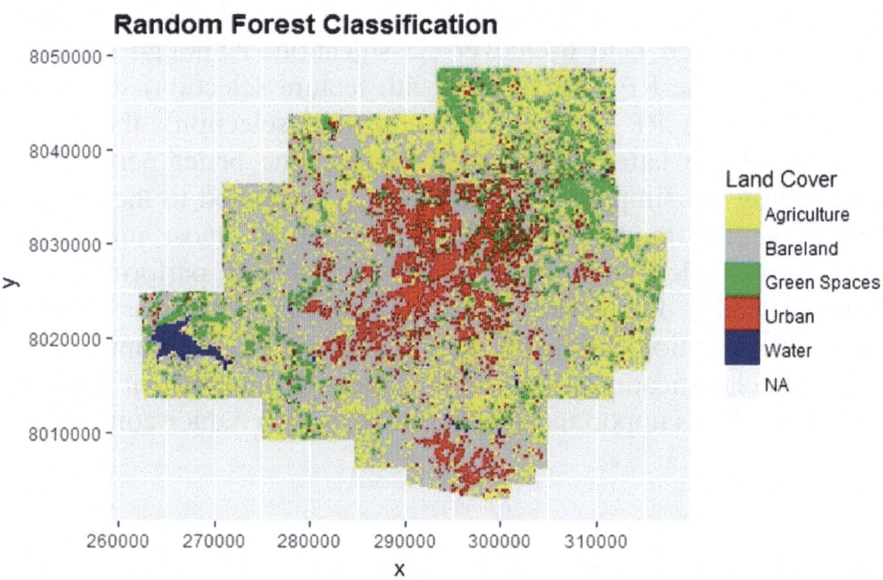

Fig. 5.8 Land cover map classified using the RFE-RF classifier (with feature selection) and multiple datasets

Finally, you can save the land use/cover map in tiff format.

```
>writeRaster(LC_rf_84,
               "RF84_MultiVariables2.tif",
                type="raw",
                datatype='INT1U',
                index=1,
                na.rm=TRUE,
                progress="window",
                overwrite=TRUE)
```

5.4.3 *Summary for Tutorial 2*

In tutorial 2, we observed that RFE-RF model did not improve land use/cover classification. However, the RFE-RF model significantly reduced the predictor variables from 42 to 28, which improved computation time of the final RF model.

5.5 Summary

We observed that the use of multiple data sets (which comprises multidate Landsat 5 TM imagery, vegetation and texture indices) slightly improved land use/cover classification. Although the OBB error rate for the RF-RFE model (with feature selection) was better than the simple RF model (without feature selection), the overall accuracy of the latter was slightly higher. The better performance achieved by the simple RF model is partly attributed to the fact that the simple RF model successfully selected the most informative predictor variables during model training. Therefore, performing RFE feature selection did not provide additional information that improves image classification. While the RFE-RF model did not improve land use/cover classification accuracy, it reduced the number of predictor variables. This is important since it can be used to reduce computation time.

5.6 Additional Exercises

i. Train RFE-SVM model and then classify the multiple data sets using the SVM classifier.
ii. Compare your results with the RFE-RF classifier.

References

Dash M, Liu H (1997) Feature selection for classification. Intell Data Anal 1(1–4):131–156

Kuhn M, Johnson K (2016) Applied predictive modeling. Springer

Yu L, Liu H (2004) Efficient feature selection via analysis of relevance and redundancy. J Mach Learn Res 5:1205–1224

Appendix

Websites with machine learning, remote sensing and GIS exercises in R.

1. **Spatial Data Science with R**

   http://rspatial.org/analysis/rst/9-remotesensing.html

2. **Stackoverflow**

   http://stackoverflow.com

3. **Burns Statistics**

   Why use the R Language?
   https://www.burns-stat.com/documents/tutorials/why-use-the-r-language/

4. **ML & R**

   Recursive Feature Elimination
   http://ml-tutorials.kyrcha.info/rfe.html

5. **Quick R by Datacamp**

   About Quick-R
   https://www.statmethods.net/

C. Kamusoko, *Remote Sensing Image Classification in R*, Springer Geography,
https://doi.org/10.1007/978-981-13-8012-9

6. **Rtips. Revival 2014!**

http://pj.freefaculty.org/R/Rtips.html

7. **Online R resources for Beginners**

http://www.introductoryr.co.uk/R_Resources_for_Beginners.html

Index

A

Absolute or physically-based calibration models, 26
Absorption, 14, 26–28
Accuracy assessment, 3, 14, 93, 106, 107, 109, 113, 120, 124, 129, 130, 139, 148, 149, 167, 178
Advanced classification methods, 2
Apparent reflectance model, 27, 42, 60
Array, 11
Artificial Neural Networks, 81–83
Atmospheric effects, 26, 27
At-satellite radiances, 27, 42
At-sensor radiance, 43, 45

B

Back-propagation, 83
Backward feature elimination, 157
Base, 6, 9, 11, 16, 19
Bootstrap sampling, 85

C

caret, 86, 93, 101, 102, 108, 144, 158
Classification, 1–5, 13–15, 25, 26, 29, 36, 61, 64, 67, 72, 74, 78, 81–84, 86, 87, 92, 93, 95, 106, 107, 109, 118, 122, 126, 130, 131, 134, 135, 148–151, 155, 156, 158, 168, 170, 171, 178–180
Classification and Regression Tree (CART), 84
Classification errors, 2, 106
Classifiers, 2, 3, 81–85, 87, 92, 93, 97, 101, 102, 122, 126, 130–133, 150, 151, 156
Clustering, 4, 5, 86
Comprehensive R Archive Network (CRAN), 6
Concatenate***c()***function, 11
confusionMatrix(), 108
Corrplot, 86, 95, 99, 141, 143, 163, 164

C. Kamusoko, *Remote Sensing Image Classification in R*, Springer Geography,
https://doi.org/10.1007/978-981-13-8012-9

Cosine Estimation of Atmospheric Transmittance (COST), 27, 28, 55
createDataPartition(), 93, 139

D
Dark Object Subtraction (DOS), 27, 42
Data-driven, 4
data.frame(), 19, 91, 137
Data frames, 11
Data type, 10
Decision boundary, 84
Deductive inference, 5
Deep learning, 2
Dense satellite imagery, 3
Digital elevation model, 156
Digital image processing, 2
Digital number, 1, 41–43
doParallel, 86, 144, 158

E
e1071, 86, 95
Earth observing remote sensing, 1
Electromagnetic radiation, 1, 43
Entropy, 69
estimateHaze(), 49, 50, 55
Euclidean distance, 82

F
Factors, 11
featurePlot(), 95, 96, 140, 141
Filter methods, 156
Forward feature selection, 157
Free and open source software systems, 2
Functions, 10

G
Generalization error, 85
Geometric correction, 25, 26
ggplot2, 28, 29, 38, 41, 45, 86, 95, 132, 140, 158
Gini impurity index, 84
Glcmpackage, 69, 73
Grey Level Co-occurrence Matrix (GLCM), 68, 69, 74, 76
grid, 86
gridExtra, 28, 29, 86, 133, 158

H
Harare, 13, 15
High data dimensionality, 155, 156
Homogeneity, 69
Hybrid methods, 2

I
Image classification, 2, 3, 67, 78, 81, 82, 85, 87, 122, 126, 133, 135, 150, 155–157, 168, 180
Image transformation, 2, 25, 26, 65
Inductive inference, 4

K
Kernel function, 84, 122
kernlab, 86

kknn, 86, 103
k-Nearest Neighbors, 81, 82

L
Landsat data, 2, 134, 150
Landsat 5 Thematic Mapper, 13
Landsat 5 TM imagery, 3, 13–15, 27, 28, 31, 34–50, 52–56, 58, 59, 62–65, 67, 69, 71, 72, 74, 78, 81, 85, 87–89, 134–136, 142, 150, 151, 155–157, 159, 167, 170, 180
Land use/cover, 1, 2, 14, 29, 67, 72, 74, 81, 82, 85, 89–91, 93, 94, 99, 104, 106, 107, 109–111, 113, 115, 117, 119, 121, 123, 126–128, 130–133, 137, 138, 146, 147, 149, 150, 166–170, 177–180

M
Machine learning, 1–5, 9, 20, 81, 82, 84–87, 92, 93, 97, 101, 122, 130–132, 144, 156, 183
maptree, 86
Matrix, 11, 19, 86, 90, 93, 99, 100, 106–109, 113, 115, 120, 121, 124, 126, 128, 129, 142, 143, 148, 163, 164, 166, 167, 177, 178
Mean, 69
Medium resolution, 2
Misclassification errors, 84, 115, 121
Moderate Resolution Atmospheric Radiance and Transmittance, 26
Modified Soil Adjusted Vegetation Index, 68
Multidate, 13, 27, 78, 81, 134–136, 143, 148–151, 155–157, 168, 170, 180
Multi-Layer Perceptron (MPL), 83

N
Near Infrared (NIR), 68
NeuralNetTools, 86, 112
Neural networks, 2, 3, 86
nnet, 86, 110
Normal distribution, 41, 83–85, 101
Normalized difference built-up index, 2
Normalized difference vegetation index, 20, 68

O
Object-based classifications, 2
olden(), 113
Outliers, 84, 85, 93, 95–97, 101, 109, 119, 122, 126, 130, 133
Out-of-Bag (OOB), 85, 128, 148, 166, 178

Overall accuracy, 104, 107, 109, 111, 112, 115, 117, 121, 124, 126–128, 130, 146, 166, 168, 177, 178, 180

P

Pattern recognition, 4
Predictor variables, 82, 85, 91, 92, 99, 102, 105, 113, 119, 122, 127, 128, 138, 143, 148, 155–159, 161, 163, 164, 166, 167, 171, 174–178, 180
Pre-processing, 2, 3, 20, 25, 29, 65, 86, 95, 104, 111, 117, 127, 165
Producer`s accuracy, 107

R

Radiometric correction, 25, 26
randomForest, 2, 10, 81, 86, 155, 166
randomForest package, 10
raster, 28, 29, 34–37, 41, 42, 70, 73, 86–88, 135, 158, 160, 169, 171, 172
rasterVis, 86, 158
RColorBrewer, 70
Recursive feature elimination, 12, 157
Red (R), 68
Refraction, 26
Regression, 4, 5, 27, 82, 83, 86
Reinforcement learning, 5
Relative calibration models, 26
Remotely-sensed image processing, 3
Remotely-sensed images, 2, 68
Remote sensing, 1–3, 20, 25, 29, 67, 82, 83, 93, 95, 105, 107, 122, 126, 183
reshape, 86, 158
rgdal, 28, 42, 70, 73, 86
R Packages, 9
rpart, 86, 116, 118, 131
RStoolbox, 11, 27–29, 33, 34, 42, 70, 73, 86, 158
RStoolboxpackage, 11, 42
RStudio, 1, 8, 9, 16–18, 20, 29–31

S

Scattering, 25–28, 42, 50, 52, 61
Second Simulation of the Satellite Signal in the Solar Spectrum, 26
Segmentation, 2
Semi-supervised, 5
Sensor's gain and offset values, 27
set.seed(), 92, 139
Simple dark object subtraction (SDOS), 27, 42
Single Decision Trees, 81, 82
Soil Adjusted Vegetation Index, 68
sp, 28, 36, 42, 70, 73, 86
Spatial heterogeneity, 3
spectral confusion, 105, 109, 115, 124
Spectral indices, 68
Spectral responses, 2
Statistical model, 4
Structural and textural information, 2
supervised, 5, 29, 81, 82

Support vector machines, 2, 3, 81, 82, 84, 86

T
Texture indices, 68
Top-of-the-atmosphere reflectance, 45
train(), 102, 104, 110, 116
trainControl(), 101, 102, 144
Training data, 3–5, 82–84, 88, 91–93, 95–98, 101, 102, 116, 122, 136, 137, 139, 140, 159–163, 171–173

U
Unmanned aerial vehicles, 1
Unsupervised, 5, 29

V
Variance, 69, 74, 161, 167, 174
Vegetation - impervious surface - soil models, 2

W
Wrapper methods, 156

The manufacturer's authorised representative in the EU is Springer Nature Customer Service Centre GmbH, Europaplatz 3, 69115 Heidelberg, Germany. If you have any concerns regarding our products, please contact ProductSafety@springernature.com

Printed and bound by CPI Group (UK) Ltd, Croydon, CR0 4YY
25/09/2026
02222900-0003